FUNCTIONS OF A COMPLEX
VARIABLE

LIBRARY OF MATHEMATICS

edited by

WALTER LEDERMANN
D.Sc., Ph.D., F.R.S.Ed., Professor of
Mathematics, University of Sussex

FUNCTIONS
OF A COMPLEX
VARIABLE

BY

D. O. TALL

ROUTLEDGE & KEGAN PAUL
LONDON, BOSTON AND HENLEY

First published in two volumes 1970
in Great Britain by
Routledge & Kegan Paul plc
14 Leicester Square,
London WC2H 7PH,
9 Park Street
Boston, Mass. 02108, USA and
Broadway House
Newtown Road
Henley-on-Thames
Oxon RG9 1EN
© D. O. Tall, 1970
Reprinted in one volume 1977
Reprinted in 1980 and 1985

ISBN 0 7100 8655 5

Printed in Great Britain by
T J Press (Padstow) Ltd,
Padstow, Cornwall

Preface

Functions of a Complex Variable forms a sequel to *Complex Numbers* by Walter Ledermann. It contains an elementary introduction to complex differential and integral calculus.

Part I is mainly concerned with differential calculus. However certain results in the theory (for example Taylor's series) are best proved by the use of integral calculus and so contour integration is introduced at an early stage. The complex analogue of the Fundamental Theorem of Calculus (which exhibits integration and differentiation as inverse operations) is discussed and this gives a natural approach to Cauchy's Theorem. This in turn yields a proof of Taylor's Theorem and a number of remarkable corollaries. For example one result states that if a complex function is assumed differentiable just once everywhere in its domain of definition, then all the higher derivatives automatically exist.

Part II contains applications of the theory, including conformal mappings, harmonic functions, calculation of integrals by residues, together with a description of analytic continuation and Riemann surfaces.

The main difficulties encountered in the text concern the theory of curves, since the sophisticated techniques (compactness, winding number etc.) which are necessary to deal with arbitrary curves are too technical for a book of this nature. Where difficulties occur, they are clearly stated. The general theory is illustrated by examples and each chapter ends with a set of exercises for the reader.

I should like to thank my colleagues Professor Walter Ledermann and Dr. Alan Weir for reading the manuscript and making many helpful suggestions and improvements in the text.

The University of Warwick   DAVID TALL

Contents

Part I

CONTENTS

Part II

1. Conformal Mappings and Harmonic Functions

2. Cauchy's Residue Theorem

3. The Calculus of Residues

4. Analytic Continuation and Riemann Surfaces

CHAPTER ONE

Differentiation

1. Preliminaries

We begin with an informal discussion on functions as a prelude to more precise assumptions which will be explained in the next section.

The notion of a function of a complex variable is intuitively very clear. Given a complex number z, there is defined uniquely another complex number $f(z)$. This is usually given by a formula, for example $f(z) = z^2$ or $f(z) = e^z$.

We also wish to consider such formulae as $f(z) = 1/z$ or the power series $f(z) = 1 + z + z^2 \ldots$ to give functions. These differ from the preceding examples in that they are not defined for every value of z. The formula $f(z) = 1/z$ is not defined for $z = 0$ and the power series is not convergent (and hence the sum is not defined) for $|z| > 1$. However they have in common with $f(z) = z^2$ and $f(z) = e^z$ the property that, for any given value of z, if $f(z)$ is defined then $f(z)$ is unique. This is not so for the expression $z^{\frac{1}{2}}$ which has two values for $z \neq 0$ or for log z which† has many values for $z \neq 0$. Such expressions are sometimes called 'many-valued functions' and they will be considered separately later. In the remainder of the text a function will always be assumed to be single-valued wherever it is defined.

If we write $z = x + iy$ where x, y are real and $f(z) = u + iv$ where u, v are real, then u, v are real functions of x, y. For this reason we write $f(z) = u(x, y) + iv(x, y)$ to illustrate that u, v depend on x, y.

† W. Ledermann, *Complex Numbers*, in this series, p. 57.

EXAMPLE 1. If $f(z) = z^2$, then $f(z) = (x+iy)^2$
$= x^2 - y^2 + 2ixy$ and so $u(x, y) = x^2 - y^2$, $v(x, y) = 2xy$.

EXAMPLE 2. If $f(z) = e^z$, then† $f(z) = e^{x+iy}$
$= e^x(\cos y + i \sin y)$ and so $u(x, y) = e^x \cos y$, $v(x, y) = e^x \sin y$.

2. The Domain of Definition of a Function

As we have seen, we wish to consider functions which are not defined everywhere. The set of complex numbers where a function is defined will be called the domain of definition. We wish to put certain restrictions on this set and these ideas are discussed in this section. It is convenient to consider the situation pictorially by identifying complex numbers with points in the plane.

If z_0 is a complex number, the *ε-neighbourhood* of z_0 is the set of all points z such that $|z - z_0| < \varepsilon$ where ε is a given positive real number.

In figure 1, the ε-neighbourhood of z_0 is the set of points in the shaded disc not including the boundary.

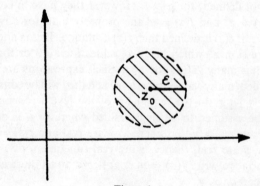

Figure 1

† W. Ledermann, *Complex Numbers*, in this series, p. 56.

A set S of points in the complex plane is said to be *open* if every point z_0 in S has an ε-neighbournood which consists entirely of points of S. For example the set C of points such that $|z| < 1$ is open, for if z_0 is in C, let $|z_0| = 1 - \delta$ where $0 < \delta \leqslant 1$, then the ε-neighbourhood of z_0 where $0 < \varepsilon \leqslant \delta$ lies completely in C.

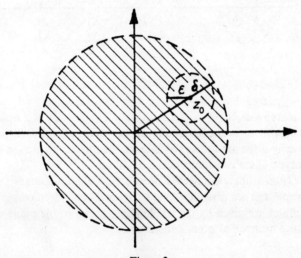

Figure 2

A *stepwise curve* in the plane is a polygonal curve, all of whose straight segments are parallel either to the real or imaginary axis.

A set S of points in the complex plane is said to be *connected* if any two points in S may be joined by a stepwise curve which lies entirely in S.

The shaded area in figure 3 is connected where z_1, z_2 are typical points.

Remark. The reader is perfectly justified in asking why we

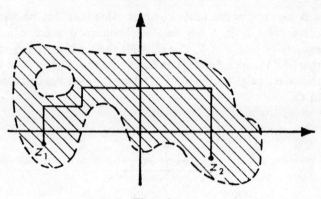

Figure 3

use a stepwise curve in the definition. The answer is simple; it is because it is the most useful (see Theorem 5.1. below). Actually if the set concerned is open, then it can be proved that any type of curve will do in the definition.

A (non-empty) connected open set is called a *domain†*. For example the set given by $|z| < 1$ (figure 2) is a domain. Other examples are given by the whole plane or the whole plane with a finite number of points missing.

Fundamental assumption. A complex function will always be assumed to be defined on a domain. This is called the *domain of definition* of the function concerned. Thus a *function of a complex variable* will be a rule which assigns to each complex number z in the domain of definition a unique complex number $f(z)$.

This rule is usually given by a 'formula' such as e^z or $1/z$ and, in common with the usual practice, we will often refer to the function by this formula.

† Some texts use the word 'region' instead of 'domain'.

Examples of functions are

  (i) e^z, defined for all z,

  (ii) $1/z$, defined for $z \neq 0$,

  (iii) $1 + z + z^2 + \dots$, defined for $|z| < 1$.

As we have remarked, $\log z$ is *not* a function in the sense that it is not single-valued. We recall that

$$\log z = \log |z| + i \,(\arg z + 2\pi k)$$

where $\log |z|$ is the usual real logarithm, $-\pi < \arg z \leqslant \pi$ and k is an integer‡. We can consider $\log z$ to be a function in the following manner: let the 'cut-plane' consist of the complex plane with the negative real axis (including zero) removed:

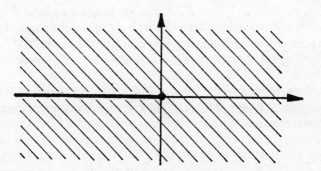

Figure 4

Now choose a fixed value of k and then $\log z$ is a single-valued function in the cut-plane. For example the principal value given by $k = 0$,

$$\mathrm{Log}\, z = \log |z| + i \arg z$$

where now $-\pi < \arg z < \pi$.

‡ W. Ledermann, *Complex Numbers*, p. 57.

Note that the cut-plane is a domain, so now $\log z$ is a function according to our definition.

In a similar manner we can consider $z^{\frac{1}{2}}$ to be a function in the cut-plane. Write $z = re^{i\theta}$ where $-\pi < \theta < \pi$ in the cut-plane, then choose the value $z^{\frac{1}{2}} = r^{\frac{1}{2}}e^{\frac{1}{2}i\theta}$. This is a function and it is usually referred to as the principal value. On the positive real axis where $\theta = 0$, it reduces to the positive square root $r^{\frac{1}{2}}$. The other value $z^{\frac{1}{2}} = r^{\frac{1}{2}}e^{i(\frac{1}{2}\theta + \pi)}$ is also a function in the cut-plane. On the positive real axis, it reduces to $r^{\frac{1}{2}}e^{i\pi} = -r^{\frac{1}{2}}$, the negative square root.

As a further example of a function defined in the cut-plane, we define the principal value of z^α to be $e^{\alpha \operatorname{Log} z}$ where α is any complex constant. Since $\operatorname{Log} z$ is uniquely defined in the cut-plane, z^α is well-defined, for example the principal value of $i^i = e^{i \operatorname{Log} i} = e^{i.i(\pi/2)} = e^{-\pi/2}$. For an integer n we have $e^{n \operatorname{Log} z} = (e^{\operatorname{Log} z})^n = z^n$ which coincides with the usual definition. For $\alpha = \frac{1}{2}$, $z = re^{i\theta}$ where $-\pi < \theta < \pi$, then $e^{\frac{1}{2}\operatorname{Log} z} = e^{\frac{1}{2}\operatorname{Log} r + \frac{1}{2}i\theta} = r^{\frac{1}{2}}e^{\frac{1}{2}i\theta}$ which corresponds to the principal value of $z^{\frac{1}{2}}$ as defined above.

Why do we insist a function is defined on a domain? Why not just on an arbitrary set? The reason will become apparent as we progress. Roughly speaking, when we discuss continuity or differentiation of a complex function at a point z_0, we would like the function to be defined near z_0 (i.e. in some ε-neighbourhood) and so require the region of definition to be open. The reason for connectedness is more subtle. It pays dividends when the function concerned is differentiable. If the function were defined on a set which consisted of several disjoint parts, the function could 'behave quite differently' on each piece. For example we could have $f(z) = z$ for $|z| < 1$, $f(z) = e^z$ for $|z| > 2$ and not defined for $1 \leqslant |z| \leqslant 2$. However if the set where the function is defined is a domain (in particular connected) and the function is differentiable, then this imposes quite strict conditions on it. Indeed, it can be shown that if we know the values of the function on part of the domain, it is determined everywhere! We leave a precise statement and proof of this remarkable result until Chapter III, but mention it to justify the introduction of a "domain of definition".

Pictorially it is impossible to represent a complex function completely and we cannot draw a graph. This is because a complex number is represented as a point in two-dimensional real space. So we would need two dimensions to represent the values of z and two for $f(z)$, making a total of four. Since we have only two-dimensional paper at our disposal, the best we can do is imagine two complex planes and as z moves about in the first, $f(z)$ moves about in the second. Of course the only values of z for which $f(z)$ is defined are those in the domain of definition so we could illustrate this by drawing the domain of definition in the first plane (denoted by D and shaded):

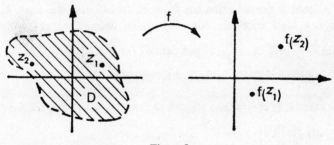

Figure 5

3. Limits and Continuity

The definitions of these concepts for complex functions are the same as in the real case†. Hence the reader familiar with the real case should find no difficulty.

DEFINITION 3.1. We say that $f(z)$ tends to the limit l as z

† P. J. Hilton, *Differential Calculus*, this series, pp. 10, 11.

tends to z_0 if the distance from $f(z)$ to l remains as small as we please so long as z remains sufficiently near to z_0, while remaining distinct from z_0.

We write $f(z) \to l$ as $z \to z_0$ or $\lim_{z \to z_0} f(z) = l$. Of course the distance from $f(z)$ to l is $|f(z) - l|$ and saying z is sufficiently near to z_0 means that $|z - z_0|$ is sufficiently small. So definition 3.1 could be phrased in terms of real numbers by defining $f(z) \to l$ as $z \to z_0$ to mean $|f(z) - l| \to 0$ as $|z - z_0| \to 0$. In precise terms, given $\varepsilon > 0$ (no matter how small), we can always find $\delta > 0$ (where δ may depend on ε) such that $0 < |z - z_0| < \delta$ implies $|f(z) - l| < \varepsilon$.

We make the usual remark that $\lim_{z \to z_0} f(z) = l$ does not mean the same as $f(z_0) = l$. The value of $f(z_0)$ is irrelevant in determining the limit because we have expressly stated in definition 3.1. that z remains distinct from z_0 in defining the limit. It is not even necessary for $f(z_0)$ to be defined, for example $\lim_{z \to 0} \dfrac{\sin z}{z} = 1$ but $\dfrac{\sin z}{z}$ is not defined for $z = 0$.

We have the usual rules for limits:

RULES. If $f(z) \to l$ and $g(z) \to k$ as $z \to z_0$, then as $z \to z_0$ we have

(i) $f(z) + g(z) \to l + k$,
(ii) $f(z) - g(z) \to l - k$,
(iii) $f(z)g(z) \to lk$,
(iv) if $k \neq 0$, $f(z)/g(z) \to l/k$.

These results can either be proved from first principles as in the real case, or by resolving each complex number into its real and imaginary parts and arguing as for limits of sequences†.

DEFINITION 3.2. We say $f(z)$ is continuous at z_0 if $f(z_0)$ is defined and $\lim_{z \to z_0} f(z)$ exists and equals $f(z_0)$.

† W. Ledermann, *Complex Numbers*, p. 47.

For example $f(z) = |z|$ is continuous everywhere. This is because $0 \leqslant ||z| - |z_0|| \leqslant |z - z_0|$, so if z is close to z_0, $|z - z_0|$ is small and $|f(z) - f(z_0)| = ||z| - |z_0||$ is small or even smaller. This shows that $f(z) \to f(z_0)$ as $z \to z_0$.

If we have a continuous function, we can imagine the situation pictorially, for if we consider a point z which approaches z_0, then the image $f(z)$ approaches $f(z_0)$.

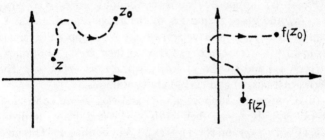

Figure 6

Continuity of a complex function is no more involved than the real case. Using rule (i) for limits, we see that if $f(z)$ and $g(z)$ are continuous at z_0, then as $z \to z_0$, we have $f(z) + g(z) \to f(z_0) + g(z_0)$ showing $f(z) + g(z)$ is continuous at z_0. Similarly the difference, product and quotient of continuous functions are continuous.

Also if $g(z)$ is continuous at z_0 and $f(w)$ is continuous at $w_0 = g(z_0)$, then $f(g(z))$ is continuous at z_0. This is because $z \to z_0$ implies $g(z) \to g(z_0) = w_0$ and so $f(g(z)) \to f(w_0) = f(g(z_0))$.

We can reduce the theory of continuity of a complex function to continuity of real functions of two real variables. Suppose $w = u + iv$, where u, v are real, then we first observe that $w \to w_0$ is equivalent to $u \to u_0$, $v \to v_0$ both together. To see this, note that

$$0 \leqslant |u - u_0| \leqslant \sqrt{\{(u - u_0)^2 + (v - v_0)^2\}} = |w - w_0|$$

and so $|u - u_0|$ is never greater than $|w - w_0|$. If $w \to w_0$ then $|w - w_0| \to 0$ implying $|u - u_0| \to 0$ and so $u \to u_0$. Similarly $v \to v_0$.

Conversely if both $u \to u_0$ and $v \to v_0$, then

$$|w - w_0| = \sqrt{\{(u - u_0)^2 + (v - v_0)^2\}} \to 0$$

and so $w \to w_0$. Using this fact we may prove:

THEOREM 3.1. If $f(z) = u(x, y) + iv(x, y)$ then the complex function $f(z)$ is continuous at $z_0 = x_0 + iy_0$ if and only if the real functions $u(x, y)$, $v(x, y)$ are continuous at (x_0, y_0).

Proof. (i) Suppose $f(z)$ is continuous at $z_0 = x_0 + iy_0$. If $x \to x_0$ and $y \to y_0$ then $z \to z_0$ by the above remark and so by continuity of $f(z)$, we have $f(z) \to f(z_0)$. Now apply the remark again to $w = f(z) = u(x, y) + iv(x, y)$ then $w \to w_0 = u(x_0, y_0) + iv(x_0, y_0)$ and so $u(x, y) \to u(x_0, y_0)$, $v(x, y) \to v(x_0, y_0)$. This shows that $u(x, y)$ and $v(x, y)$ are continuous.

(ii) Conversely, suppose $u(x, y)$ and $v(x, y)$ are continuous at (x_0, y_0). If $z \to z_0$, then both $x \to x_0$ and $y \to y_0$ implying $u(x, y) \to u(x_0, y_0)$ and $v(x, y) \to v(x_0, y_0)$ by continuity. This gives

$$u(x, y) + iv(x, y) \to u(x_0, y_0) + iv(x_0, y_0)$$

that is to say $f(z) \to f(z_0)$ and so $f(z)$ is continuous.

EXAMPLE 1. arg z is continuous in the cut-plane.

This is a basic result that we will need later and it is quite tricky to prove. The method we give uses a theorem from real variable theory†.

Suppose that $t = h(\theta)$ $(\alpha \leqslant \theta \leqslant \beta)$ is a real-valued monotonic strictly increasing function where $h(\alpha) = a$, $h(\beta) = b$, then we may solve this to find θ in terms of t, $\theta = g(t)$ $(a \leqslant t \leqslant b)$. Furthermore, if h is continuous, then so is g. For example $t = \sin \theta$ $\left(-\dfrac{\pi}{2} \leqslant \theta \leqslant \dfrac{\pi}{2} \right)$ is such a function, taking every value in $-1 \leqslant t \leqslant 1$ and so $\theta = \sin^{-1} t$ is well-defined and continuous

† Scott & Tims, *Mathematical Analysis*, Cambridge University Press, p. 217.

for $-1 \leqslant t \leqslant 1$, taking values in $-\dfrac{\pi}{2} \leqslant \sin^{-1} t \leqslant \dfrac{\pi}{2}$.

First consider the domain given by $x > 0$.

Here $\arg z = \sin^{-1}\left(\dfrac{y}{\sqrt{(x^2+y^2)}}\right)$

where we choose $-\dfrac{\pi}{2} < \sin^{-1}\left(\dfrac{y}{\sqrt{(x^2+y^2)}}\right) < \dfrac{\pi}{2}$.

But $\sqrt{(x^2+y^2)} = |z|$ is continuous and non-zero for $x > 0$, hence $y/|z|$ is a continuous function of x, y for $x > 0$. Thus $\sin^{-1}(y/|z|)$ is a continuous function of a continuous function and hence continuous. This shows that $\arg z$ is continuous in the domain $x > 0$.

Similarly we may show that $\arg z$ is continuous for $y > 0$ by considering $\arg z = \cos^{-1}\left(\dfrac{x}{\sqrt{(x^2+y^2)}}\right)$ where we choose $0 \leqslant \cos^{-1}\left(\dfrac{x}{\sqrt{(x^2+y^2)}}\right) \leqslant \pi$. (Note that cos is monotonic decreasing here.) Finally $\arg z$ is continuous for $y < 0$ where we have $\arg z = \cos^{-1}\left(\dfrac{x}{\sqrt{(x^2+y^2)}}\right)$, this time choosing $-\pi \leqslant \cos^{-1}\left(\dfrac{x}{\sqrt{(x^2+y^2)}}\right) \leqslant 0$. The three domains $x > 0$, $y > 0$, $y < 0$ together cover the cut-plane and the result is proved.

EXAMPLE 2. Log z is continuous in the cut-plane.
This follows from example 1 and theorem 3.1 because $\text{Log } z = \log |z| + i \arg z$. Note that the real part $\log |z| = \log (x^2+y^2)^{\frac{1}{2}}$ is continuous for $(x, y) \neq (0, 0)$ and the imaginary part is continuous in the cut-plane. This gives the required result.

4. Differentiation

As with limits and continuity, differentiation of a complex

function is defined in the same way as the real case.† (Notice however that the derivative is no longer the gradient of a graph because we cannot draw the graph of a complex function.)

DEFINITION 4.1. The derivative at z_0 of the function $f(z)$ is

$$f'(z_0) = \lim_{z \to z_0} \frac{f(z) - f(z_0)}{z - z_0}.$$

If we make a change of variable $z - z_0 = h$ we also have

$$f'(z_0) = \lim_{h \to 0} \frac{f(z_0 + h) - f(z_0)}{h}.$$

EXAMPLE. $f(z) = z^2$

$$f'(z_0) = \lim_{h \to 0} \frac{(z_0 + h)^2 - z_0^2}{h} = \lim_{h \to 0} (2z_0 + h) = 2z_0.$$

Often we use the alternative notation $w = f(z)$ and $\dfrac{dw}{dz} = f'(z)$.

Of course $f'(z_0)$ may not exist. As a trivial example, if $f(0) = 0$ and $f(z) = 1$ for $z \neq 0$, then $f'(0)$ does not exist. For $h \neq 0$, $\dfrac{f(h) - f(0)}{h} = \dfrac{1}{h}$ and this does not tend to a finite limit as $h \to 0$. We now show that a differentiable function is necessarily continuous (thus we may infer that a discontinuous function cannot be differentiable).

THEOREM 4.1. If $f(z)$ is differentiable at z_0, then $f(z)$ is continuous at z_0.

Proof. $\lim\limits_{z \to z_0} (f(z) - f(z_0)) = \lim\limits_{z \to z_0} \dfrac{(f(z) - f(z_0))}{z - z_0} (z - z_0)$

$= \lim\limits_{z \to z_0} \dfrac{f(z) - f(z_0)}{z - z_0} \lim\limits_{z \to z_0} (z - z_0)$ by the rule for limits

† P. J. Hilton, *Differential Calculus*, p. 12.

$$= f'(z_0).0$$
$$= 0.$$

So $\lim_{z \to z_0} f(z) = f(z_0)$ and $f(z)$ is continuous at z_0.

We may verify the usual rules for differentiation as in the real case†:

RULES.

(i) $\dfrac{d}{dz}(Af(z)+Bg(z)) = Af'(z)+Bg'(z)$ where A, B are (complex) constants.

(ii) $\dfrac{d}{dz}(f(z)\,g(z)) = f(z)\,g'(z)+f'(z)\,g(z)$

(iii) $\dfrac{d}{dz}(f(z)/g(z)) = \{f'(z)\,g(z)-f(z)\,g'(z)\}/(g(z))^2$ if $g(z) \neq 0$

(iv) $\dfrac{d}{dz}f(g(z)) = f'(g(z))\,g'(z)$.

EXAMPLE. If n is an integer, $\dfrac{d}{dz}(z^n) = nz^{n-1}$.

Proof by induction. For $n = 1$, $\dfrac{d}{dz}(z) = \lim_{h \to 0} \dfrac{(z+h)-z}{h} = 1$

and so the formula is true.

Assume it for n, then $\dfrac{d}{dz}(z^{n+1}) = \dfrac{d}{dz}(z^n z)$

$$= z^n \frac{d}{dz}(z)+\frac{d}{dz}(z^n)z \text{ by (ii)}$$

$$= z^n+nz^{n-1}z$$

$$= (n+1)z^n \text{ and so the formula is true for } n+1 \text{ and by}$$
induction true for all positive integers.

Using (i) and (iv), we may calculate the derivative of a rational

† P. J. Hilton, *Differential Calculus*, pp. 16–19.

function $\dfrac{a_n z^n + \ldots + a_0}{b_m z^m + \ldots + b_0}$ in the same way, for example

$$\frac{d}{dz}\frac{z+1}{z^2+2} = \left\{\frac{d}{dz}(z+1).(z^2+2) - (z+1)\frac{d}{dz}(z^2+2)\right\}\Big/(z^2+2)^2$$

$$= (z^2+2-2z^2-2z)/(z^2+2)^2$$

$$= (2-2z-z^2)/(z^2+2)^2$$

A rational function is differentiable wherever it is defined (i.e. whenever $b_m z^m + \ldots + b_0 \neq 0$).

DEFINITION 4.2. A function of a complex variable is said to be *analytic*† if it is differentiable everywhere in its domain of definition.

For example rational functions are analytic.

5. The Cauchy-Riemann Equations

We now come across the first property that distinguishes the complex theory from the real. When calculating $f'(z_0) = \lim\limits_{z \to z_0} \dfrac{f(z_0)-f(z)}{z-z_0}$, we may let z approach z_0 in *any* fashion.

Let us calculate $f'(z_0)$ in two distinct ways:

(i) Let $z_0 = x_0+iy_0$, $z = x_0+h+iy_0$ where h is real, and write $f(z) = u(x,y)+iv(x,y)$ then $f'(z_0) = \lim\limits_{z \to z_0} \dfrac{f(z)-f(z_0)}{z-z_0}$

$$= \lim_{h \to 0}\left\{\frac{f((x_0+h)+iy_0)-f(x_0+iy_0)}{h}\right\}$$

$$= \lim_{h \to 0}\left\{\frac{u(x_0+h,y_0)+iv(x_0+h,y_0)-u(x_0,y_0)-iv(x_0,y_0)}{h}\right\}$$

$$= \lim_{h \to 0}\left\{\frac{u(x_0+h,y_0)-u(x_0,y_0)}{h}+\frac{i(v(x_0+h,y_0)-v(x_0,y_0))}{h}\right\}$$

† Some texts use the word 'regular' instead of 'analytic'.

$$= \frac{\partial u}{\partial x} + i\frac{\partial v}{\partial x}.$$

(ii) Let $z_0 = x_0 + iy_0$, $z = x_0 + i(y_0 + k)$ where k is real, then as $z \to z_0$, we have $k \to 0$ and so

$f'(z_0)$
$$= \lim_{z \to z_0} \frac{f(z) - f(z_0)}{z - z_0}$$
$$= \lim_{k \to 0} \frac{f(x_0 + i(y_0 + k)) - f(x_0 + iy_0)}{ik}$$
$$= \lim_{k \to 0} \left\{ \frac{u(x_0, y_0 + k) + iv(x_0, y_0 + k) - u(x_0, y_0) - iv(x_0, y_0)}{ik} \right\}$$
$$= \lim_{k \to 0} \left\{ \frac{v(x_0, y_0 + k) - v(x_0, y_0)}{k} - \frac{i(u(x_0, y_0 + k) - u(x_0, y_0))}{k} \right\}$$
$$= \frac{\partial v}{\partial y} - i\frac{\partial u}{\partial y}.$$

Since $f'(z_0)$ is uniquely defined no matter how we let z approach z_0, we must have

$$f'(z_0) = \frac{\partial u}{\partial x} + i\frac{\partial v}{\partial x} = \frac{\partial v}{\partial y} - i\frac{\partial u}{\partial y}.$$

Comparing real and imaginary parts, we find that

$$\frac{\partial u}{\partial x} = \frac{\partial v}{\partial y}, \ \frac{\partial v}{\partial x} = -\frac{\partial u}{\partial y}.$$

These are called the Cauchy-Riemann equations which hold for differentiable complex functions. They give a simple way of asserting a function is *not* differentiable.

EXAMPLE. $f(z) = |z|$. Here $u(x, y) = \sqrt{(x^2 + y^2)}$, $v(x, y) = 0$, giving:

$$\frac{\partial u}{\partial x} = \frac{2x}{\sqrt{(x^2 + y^2)}}, \ \frac{\partial u}{\partial y} = \frac{2y}{\sqrt{(x^2 + y^2)}}, \ \frac{\partial v}{\partial x} = 0 = \frac{\partial v}{\partial y}.$$

Hence if either $x \neq 0$ or $y \neq 0$, at least one of the equations $\frac{\partial u}{\partial x} = \frac{\partial v}{\partial y}$, $\frac{\partial v}{\partial x} = -\frac{\partial u}{\partial y}$ does not hold. If both $x = 0$ and $y = 0$, then substituting in $\frac{\partial u}{\partial x}$, $\frac{\partial u}{\partial y}$ we get $\frac{0}{0}$. Returning to first principles

$$\frac{\partial u}{\partial x}(0, 0) = \lim_{k \to 0} \frac{u(k, 0) - u(0, 0)}{k} = \lim_{k \to 0} \frac{\sqrt{k^2}}{k}$$
$$= \lim_{k \to 0} \frac{|k|}{k}.$$

Since

$$\frac{|k|}{k} = \begin{cases} 1 \text{ for } k > 0 \\ -1 \text{ for } k < 0, \end{cases}$$

the limit of $\frac{|k|}{k}$ as $k \to 0$ does not exist and so $\frac{\partial u}{\partial x}(0, 0)$ is not defined. Similarly $\frac{\partial v}{\partial x}(0, 0)$ does not exist and the Cauchy-Riemann equations cannot hold at the origin.

Thus $f(z) = |z|$ is not differentiable anywhere but it is continuous everywhere!

The reader who may be upset by this seemingly unnatural state of affairs may be consoled by the fact that as well as z^n, the standard functions, e^z, cos z, sin z etc. are all analytic. As a possible proof of this fact we might use the Cauchy-Riemann equations. This meets with a small obstacle.

If $f(z) = u(x, y) + iv(x, y)$ is a complex function such that the partial derivatives $\frac{\partial u}{\partial x}$, $\frac{\partial v}{\partial x}$, $\frac{\partial u}{\partial y}$, $\frac{\partial v}{\partial y}$ exist and satisfy the Cauchy-Riemann equations, then $f(z)$ need not be differentiable. For example the (rather synthetic) function given by $f(z) = 1$ if $x = 0$ or $y = 0$, but $f(z) = 0$ otherwise satisfies the Cauchy-Riemann equations at the origin (all partial derivatives are zero) but it is not differentiable there because it is not even continuous.

However if the partial derivatives $\dfrac{\partial u}{\partial x}$, $\dfrac{\partial v}{\partial x}$, $\dfrac{\partial u}{\partial y}$, $\dfrac{\partial v}{\partial y}$ exist, satisfy the Cauchy-Riemann equations and are all *continuous*, then we *can* infer that the function is differentiable. The proof of this fact will be omitted since we will not use the result later. However, as an example of its possible use, consider $f(z) = e^z = e^{x+iy} = e^x \cos y + ie^x \sin y$.

$$\frac{\partial u}{\partial x} = e^x \cos y = \frac{\partial v}{\partial y}, \quad \frac{\partial v}{\partial x} = e^x \sin y = -\frac{\partial u}{\partial y}$$

and all the partial derivatives are continuous, so by the above remark, e^z is differentiable with derivative

$$\frac{\partial u}{\partial x} + i\frac{\partial v}{\partial x} = \frac{\partial v}{\partial y} - i\frac{\partial u}{\partial y} = e^x \cos y + ie^x \sin y = e^z.$$

Thus we have verified the equation $\dfrac{d(e^z)}{dz} = e^z$, already well-known in its real form. In the next section we will demonstrate this in a different way using the power series expansion for e^z.

We close this section with the following:

THEOREM 5.1. If $f(z)$ is a function of a complex variable defined on a domain D, then $f'(z) = 0$ for all z in D implies $f(z)$ is constant.

Remark. If $f'(z) = 0$ and the function were defined on a set that was not connected, this theorem need not be true. The function could be constant on each connected piece, but the constants need not be the same. For example on the set $|z| < 1$ or $|z| > 2$, define $f(z) = 0$ for $|z| < 1$ and $f(z) = 1$ for $|z| > 2$ then $f'(z) = 0$, but $f(z)$ is not constant.

Proof of theorem. If $f'(z) = 0$ then

$$\frac{\partial u}{\partial x} + i\frac{\partial v}{\partial x} = \frac{\partial v}{\partial y} - i\frac{\partial u}{\partial y} = 0$$

implying

$$\frac{\partial u}{\partial x} = \frac{\partial v}{\partial x} = \frac{\partial u}{\partial y} = \frac{\partial v}{\partial y} = 0.$$

Now $\frac{\partial u}{\partial x} = 0$ i.e. $\frac{\partial u}{\partial x}(x, y_0) = 0$ for fixed y_0. From real variable theory[†] we have $u(x, y_0) = $ constant (since $u(x, y_0)$ is a real function of the real variable x). This means $u(x, y)$ is constant along any horizontal line segment $y = y_0$ ($=$ constant) in the domain of definition. Similarly $\frac{\partial u}{\partial y} = 0$ implies that $u(x, y)$ is constant along any vertical line segment $x = x_0$ ($=$ constant) in the domain of definition. Also from $\frac{\partial v}{\partial x} = \frac{\partial v}{\partial y} = 0$ we reach the same conclusions for $v(x, y)$. Hence $f(z) = u(x, y) + iv(x, y)$ is constant along each horizontal or vertical segment in the domain of definition.

But a domain is *connected*, (here comes the full force of the definition), and any two points z_1, z_2 in the domain of definition may be joined by a stepwise curve which lies entirely in the domain (refer back to figure 3). Since $f(z)$ is constant along each segment of this curve, we must have $f(z_1) = f(z_2)$. Since z_1, z_2 are arbitrary points in the domain, $f(z)$ is constant.

6. Power Series

We assume the reader has already met the idea of a power series

$$c_0 + c_1 z + c_2 z^2 + \ldots + c_n z^n + \ldots$$

where z is a complex variable and $c_0, c_1, c_2, \ldots$ are fixed complex numbers[‡]. We recall that either there is a positive number R (called the radius of convergence) such that the series

---

[†] P. J. Hilton, *Differential Calculus*, p. 37.
[‡] W. Ledermann, *Complex Numbers*, p. 49.

converges absolutely for $|z| < R$ and diverges for $|z| > R$, or the series converges absolutely for all z (in which case we formally write $R = \infty$)[†]. Thus in our terminology the power series is a function with domain $|z| < R$.

EXAMPLE (i). $1 + z + z^2 + \ldots + z^n + \ldots$ has radius of convergence $R = 1$, and for $|z| < 1$, we have

$$1 + z + z^2 + \ldots + z^n + \ldots = (1 - z)^{-1}.$$

EXAMPLES (ii)—(iv) all have infinite radius of convergence.

(ii) $e^z = 1 + \dfrac{z}{1!} + \dfrac{z^2}{2!} + \ldots + \dfrac{z^n}{n!} + \ldots$

(iii) $\sin z = z - \dfrac{z^3}{3!} + \dfrac{z^5}{5!} - \ldots$

(iv) $\cos z = 1 - \dfrac{z^2}{2!} + \dfrac{z^4}{4!} - \ldots$

It has already been noted that inside the circle of convergence power series may be manipulated in much the same way as polynomials[‡]. For example two power series may be added term by term. The same is true of differentiation. A power series may be differentiated term by term inside the circle of convergence and if $f(z) = c_0 + c_1 z + c_2 z^2 + \ldots + c_n z^n + \ldots$ for $|z| < R$, then $f'(z) = c_1 + 2c_2 z + \ldots + nc_n z^{n-1} + \ldots$ for $|z| < R$. A proof of this result is somewhat technical and may be found in Appendix I. Note that a power series is differentiable everywhere in its domain of definition and so it is an analytic function.

EXAMPLE (i). $f(z) = 1 + z + z^2 + \ldots + z^n + \ldots$ $|z| < 1$, then $f'(z) = 1 + 2z + \ldots + nz^{n-1} + \ldots$ $|z| < 1$.

[†] W. Ledermann, *Complex Numbers*, p. 50.
[‡] W. Ledermann, *Complex Numbers*, p. 51.

Since $f(z) = (1-z)^{-1}$ in this case, differentiating we have $f'(z) = (1-z)^{-2}$ and so

$$(1-z)^{-2} = 1+2z+ \ldots +nz^{n-1}+ \ldots \quad |z| < 1.$$

(ii) $e^z = 1+\dfrac{z}{1!}+\dfrac{z^2}{2!}+ \ldots +\dfrac{z^n}{n!}+ \ldots$

$$\frac{d}{dz}(e^z) = e^z.$$

(iii) $\sin z = z-\dfrac{z^3}{3!}+\dfrac{z^5}{5!}- \ldots \ldots$

$$\frac{d}{dz}(\sin z) = \cos z.$$

Similarly (iv) $\dfrac{d}{dz}(\cos z) = -\sin z.$

EXAMPLE (v). Since $w = \operatorname{Log} z$ is defined by the equation $z = e^w$, we can use this to show $\dfrac{dw}{dz} = \dfrac{1}{z}$. The function $w = \operatorname{Log} z$ is continuous in the cut-plane and so as $z \to z_0$, we have $w \to w_0$.

$$\frac{dw}{dz} = \lim_{z \to z_0} \frac{\operatorname{Log} z - \operatorname{Log} z_0}{z - z_0}$$

$$= \lim_{w \to w_0} \frac{w - w_0}{e^w - e^{w_0}}$$

$$= \lim_{w \to w_0} \left\{ \frac{e^w - e^{w_0}}{w - w_0} \right\}^{-1}$$

$$= 1 \left/ \frac{d}{dw}(e^w) \right.$$

$$= 1/e^w$$

$$= 1/z.$$

EXAMPLE (vi). $\frac{d}{dz}(z^{\alpha}) = \alpha z^{\alpha-1}$ in the cut-plane.

This is because $z^{\alpha} = e^{\alpha \operatorname{Log} z} = f(g(z))$ where $f(w) = e^w$, $g(z) = \alpha \operatorname{Log} z$ and so $\frac{d}{dz}(z^{\alpha}) = f'(g(z))g'(z) = e^{\alpha \operatorname{Log} z} \cdot \frac{\alpha}{z} = z^{\alpha} \cdot \frac{\alpha}{z} = \alpha z^{\alpha-1}$.

Notice that the derivative of a power series is again a power series with the same domain of definition. This means we can differentiate it again, indeed we can differentiate it as many times as we like, so that if

$$f(z) = c_0 + c_1 z + c_2 z^2 + c_3 z^3 + \ldots + c_n z^n + \ldots$$

then

$$f'(z) = c_1 + 2c_2 z + 3c_3 z^2 + 4c_4 z^3 + \ldots + nc_n z^{n-1} + \ldots$$
$$f''(z) = 2c_2 + 6c_3 z + 12c_4 z^2 + \ldots + n(n-1)c_n z^{n-2} + \ldots$$
$$f'''(z) = 6c_3 + 24c_4 z + \ldots + n(n-1)(n-2)c_n z^{n-3} + \ldots$$
etc.

Putting $z = 0$ in these equations we find $f(0) = c_0$, $f'(0) = c_1 = 1!c_1$, $f''(0) = 2c_2 = 2!c_2$, $f'''(0) = 6c_3 = 3!c_3$ and in general $f^{(n)}(0) = n!c_n$. This means that by substituting these values in the series we may write it in its usual Taylor-MacLaurin form:

$$f(z) = c_0 + c_1 z + c_2 z^2 + \ldots + c_n z^n + \ldots$$
$$= f(0) + \frac{f'(0)}{1!}z + \frac{f''(0)}{2!}z^2 + \ldots + \frac{f^{(n)}(0)}{n!}z^n + \ldots$$

More generally we may consider a power series centred on z_0:

$$f(z) = a_0 + a_1(z - z_0) + \ldots + a_n(z - z_0)^n + \ldots \text{ for } |z - z_0| < R.$$

In this case we have:

$$f'(z) = a_1 + 2a_2(z - z_0) + \ldots + na_n(z - z_0)^{n-1} + \ldots \quad |z - z_0| < R.$$
etc.

and we find $f^{(n)}(z_0) = n!a_n$.

This gives:

PROPOSITION 6.1. If $f(z) = a_0 + a_1(z - z_0) + \ldots +$ $a_n(z - z_0)^n + \ldots$ for $|z - z_0| < R$, then f is differentiable as many times as we please for $|z - z_0| < R$ and $a_n = \dfrac{f^{(n)}(z_0)}{n!}$.

We remark at this stage that power series seem to be very special. They are not only differentiable, we can differentiate as many times as we please. In the real theory it is possible to invent functions which are differentiable once, but not twice. (e.g. $f(x) = 0$ for $x \leqslant 0$ and $f(x) = x^2$ for $x \geqslant 0$. Here $f'(x) = 0$ for $x \leqslant 0$ and $f'(x) = 2x$ for $x \geqslant 0$. Note that $f'(0) = 0$ calculated from either side. However $f''(0)$ does not exist, being 0 calculated from the left and 2 calculated from the right.)

It is a very pleasant (and surprising) fact that in the complex theory, if a function is analytic (i.e. differentiable once everywhere in its domain of definition) then it is differentiable as many times as we please. We will demonstrate this fact later (pages 55–56). How is it proved?—By using power series.

EXERCISES ON CHAPTER ONE

1. Write $f(z) = u(x, y) + iv(x, y)$ and find $u(x, y)$, $v(x, y)$ in each of the following cases:
   (i) $z^2 + 2z$ (ii) $1/z$ $(z \neq 0)$ (iii) $\sin z$ (iv) $z/(e^z - 1)$ $(z \neq 0)$
   (v) Log z in the cut-plane   (vi) $|z|^2$ (vii) arg z in the cut-plane.

2. In exercise 1, differentiate (i)–(v).

3. (a) In exercise 1 (vi), calculate $\dfrac{\partial u}{\partial x}, \dfrac{\partial u}{\partial y}, \dfrac{\partial v}{\partial x}, \dfrac{\partial v}{\partial y}$ for $(x, y) \neq$ (0, 0). Hence show that $|z|^2$ is not differentiable for $z \neq 0$. What happens at $z = 0$? (Hint: use first principles).

(b) In 1 (vii), calculate $\dfrac{\partial u}{\partial x}, \dfrac{\partial u}{\partial y}, \dfrac{\partial v}{\partial x}, \dfrac{\partial v}{\partial y}$ and hence show that arg z is not differentiable anywhere.

4. In each of the following cases, draw a sketch of the given set and say if it is a domain where z is subject to the given restriction:
   (i) $|z-1| < 2$ (ii) $|z| \leqslant 1$ (iii) $x < -1$ or $x > 1$ where $z = x + iy$
   (iv) $z \neq t$ where t is real and $t \leqslant 0$ (v) $1 < |z| < 2$.

5. Suppose that f is an analytic function such that $f(z)$ is always real. Use the Cauchy-Riemann equations to prove that f is constant.

6. (i) Substitute $w = e^{\lambda z}$ in the equation $\dfrac{d^2w}{dz^2} + k^2w = 0$ $(k \neq 0)$ to find λ_1, λ_2 such that $e^{\lambda_1 z}$ and $e^{\lambda_2 z}$ are solutions. Show that $Ae^{\lambda_1 z} + Be^{\lambda_2 z}$ is also a solution where A, B are complex constants.

   Use the same method to find solutions for

   (ii) $\dfrac{d^2w}{dz^2} - \dfrac{3dw}{dz} + 2w = 0$ (iii) $\dfrac{d^3w}{dz^3} - \dfrac{d^2w}{dz^2} + \dfrac{4dw}{dz} - 4w = 0$.

7. Use $(1-z)^{-2} = 1 + 2z + 3z^2 + \ldots + nz^{n-1} + \ldots$ $|z| < 1$ to find, by differentiation, a power series formula for $(1-z)^{-4}$ valid for $|z| < 1$.

8. Consider $f(z) = 1 + \displaystyle\sum_{n-1}^{\infty} \dfrac{\alpha(\alpha-1) \ldots\ldots (\alpha-n+1)}{n!} z^n$

   Prove that the series is absolutely convergent for $|z| < 1$ and that $f'(z) = \alpha f(z)/(1+z)$. Consider $\phi(z) = \dfrac{f(z)}{(1+z)^\alpha}$ and show that $\phi'(z) = 0$ for $|z| < 1$. Hence conclude that $f(z) = (1+z)^\alpha$ for $|z| < 1$.

9. Show $f(z) = z - \dfrac{z^2}{2} + \dfrac{z^3}{3} - \dfrac{z^4}{4} + \ldots\ldots$ is absolutely convergent for $|z| < 1$ and that $f'(z) = (1+z)^{-1}$.
   Hence conclude that $f(z) = \text{Log}\,(1+z)$ for $|z| < 1$.

CHAPTER TWO

Integration

---

1. Contours

If $\phi(t)$, $\psi(t)$ are continuous real functions of the real variable t defined in an interval $\alpha \leqslant t \leqslant \beta$, the equation

$$z(t) = \phi(t) + i\psi(t) \quad (\alpha \leqslant t \leqslant \beta) \tag{1}$$

determines a *path* in the complex plane†. Thus a path is a continuous function defined on the interval $\alpha \leqslant t \leqslant \beta$, taking values in the complex plane. The *initial* and *final* points are $z(\alpha)$, $z(\beta)$ respectively and the path is said to be *closed* if $z(\alpha) = z(\beta)$. As t increases, the point $z(t)$ traverses a curve in the complex plane from $z(\alpha)$ to $z(\beta)$.

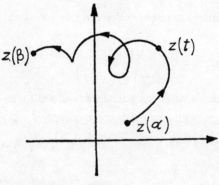

Figure 7

† This is the same definition as a path in the (x, y)-plane given in *Multiple Integrals* by W. Ledermann, p. 1.

The set of points in the complex plane given by $z(t)$ for $\alpha \leqslant t \leqslant \beta$ is called the *track*. Two different paths may have the same track, for example the paths

$$z_1(t) = \cos t + i \sin t \qquad \left(0 \leqslant t \leqslant \frac{\pi}{2} \right) \qquad (2)$$

$$z_2(t) = \frac{1-t^2}{1+t^2} + \frac{2it}{1+t^2} \qquad (0 \leqslant t \leqslant 1) \qquad (3)$$

both determine the track in figure 8:

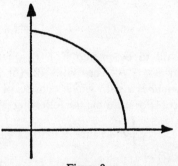

Figure 8

For this reason, if we refer to a pictorial representation of a curve and wish to talk about the path, then we should also specify the function which determines it. For example we will choose the standard function which represents the unit circle as a path to be

$$z(t) = \cos t + i \sin t \qquad (0 \leqslant t \leqslant 2\pi). \qquad (4)$$

More generally, the circle centre z_0, radius r will be

$$z(t) = z_0 + re^{it} \qquad (0 \leqslant t \leqslant 2\pi). \qquad (5)$$

The line segment from z_1 to z_2 will be

$$z(t) = z_1(1-t) + z_2 t \qquad (0 \leqslant t \leqslant 1). \qquad (6)$$

We sometimes refer to the path as a 'parametrization' of the track and call t the 'parameter'.

The *opposite path* to (1) is the path

$$z_0(t) = z(\alpha + \beta - t)$$
$$= \phi(\alpha + \beta - t) + i\psi(\alpha + \beta - t) \qquad (\alpha \leqslant t \leqslant \beta). \qquad (7)$$

Notice that as t increases, $z_0(t)$ traverses the same track as $z(t)$, but in the opposite sense. For example the opposite path to (2) is

$$z_0(t) = \cos\left(\frac{\pi}{2} - t\right) + i \sin\left(\frac{\pi}{2} - t\right)$$
$$= \sin t + i \cos t \qquad \left(0 \leqslant t \leqslant \frac{\pi}{2}\right) \qquad (8)$$

A path is said to be *smooth* if $\phi'(t)$, $\psi'(t)$ exist and are continuous for $\alpha \leqslant t \leqslant \beta$. The paths (2)–(6) and (8) are all smooth. A *contour* is a path which consists of a finite number of smooth pieces. For example the following path is a contour:

$$z(t) = \begin{cases} t^2 + i \sin \dfrac{\pi}{2} t & (0 \leqslant t \leqslant 1) \\[2mm] t + it & (1 \leqslant t \leqslant 2) \\[2mm] 4 - t + 2i & (2 \leqslant t \leqslant 3). \end{cases}$$

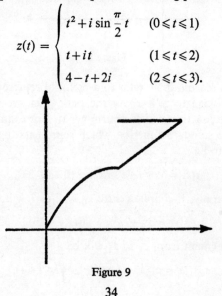

Figure 9

A *Jordan contour* is a contour

$$z(t) = \phi(t) + i\psi(t) \qquad (\alpha \leqslant t \leqslant \beta)$$

such that $t_1 \neq t_2$ implies $z(t_1) \neq z(t_2)$. Thus a Jordan contour has no self-intersections. A *closed Jordan contour* is a closed contour such that $\alpha \leqslant t_1 < t_2 < \beta$ implies $z(t_1) \neq z(t_2)$. In this case there are no self-intersections other than the coincident endpoints. An example is given by the unit circle

$$z(t) = \cos t + i \sin t \qquad (0 \leqslant t \leqslant 2\pi).$$

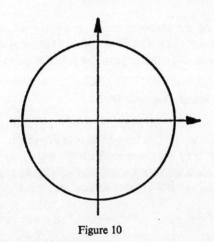

Figure 10

The *Jordan curve theorem* states that a closed Jordan contour divides the plane into two domains, one bounded (called the interior) and one unbounded (the exterior).

The reader who relies on his geometric intuition may feel that this result is patently obvious. For example the interior of the unit circle in figure 10 is certainly given by $|z| < 1$ and the exterior by $|z| > 1$. For every particular Jordan contour we meet in this text, the result will be clear. However it is possible to draw 'maze-like' Jordan curves such as:

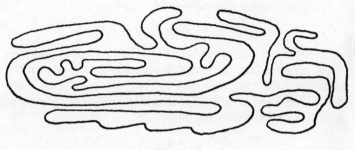

Figure 11

A proof of the theorem must be written so as to include every possiblity which may arise and it is not surprising that this is very difficult. For this reason the proof is omitted.

2. Contour Integration

As in the real case, we wish to discuss integration of a complex function. There is no immediate analogue to the symbol $\int_{z_1}^{z_2} f(z)\, dz$ where f is a complex function and z_1, z_2 are complex numbers. This is because we may consider z_1, z_2 as points in the plane and the symbol $\int_{z_1}^{z_2} f(z)\, dz$ does not specify how z varies between z_1 and z_2. To do this, the integral is defined along a contour γ from z_1 to z_2. In general this will depend on the choice of γ and so we use the notation $\int_\gamma f(z) dz$.

Note that we require f to be defined everywhere on the track of γ. Equivalently, if f is defined in the domain D, we could ask that γ lies in D (meaning, of course, that the track of γ lies in D). This is to be preferred, because then we can visualize a picture of the domain with (the track of) γ lying in it.

First we define an integral along a smooth path. Let f be a continuous complex function defined in a domain D and suppose that γ is the smooth path $z(t) = x(t) + i y(t)$ $(\alpha \leqslant t \leqslant \beta)$, where γ lies in D. Since γ is smooth, $z'(t) = x'(t) + i y'(t)$ exists and is continuous for $\alpha \leqslant t \leqslant \beta$.

36

Define

$$\int_\gamma f(z)\,dz = \int_\alpha^\beta f(z)\,\frac{dz}{dt}\,dt \tag{1}$$

By writing $f(z) = u(x,\,y) + iv(xy)$, we have

$$f(z(t)) = u(x(t),\,y(t)) + iv(x(t),\,y(t))$$
$$= \underline{u}(t) + i\underline{v}(t),$$

and so equation (1) becomes

$$\int_\gamma f(z)\,dz = \int_\alpha^\beta (\underline{u}(t) + i\underline{v}(t))\,(x'(t) + iy'(t))dt \tag{2}$$

i.e. $$\int_\gamma f(z)\,dz = \int_\alpha^\beta (\underline{u}x' - \underline{v}y')dt + i\int_\alpha^\beta (\underline{u}y' + \underline{v}x')\,dt \tag{3}$$

Equation (3) states that to integrate along a smooth path, we substitute in terms of the parameter t, separate into real and imaginary parts, and calculate two real integrals.

EXAMPLE 1. Integrate z^2 along the smooth path γ given by $z(t) = t + it^2$ $(0 \leqslant t \leqslant 1)$.

Since $z'(t) = 1 + 2it$, we have

$$\int_\gamma z^2 dz = \int_0^1 (t + it^2)^2\,(1 + 2it)dt$$
$$= \int_0^1 (t^2 - 5t^4)dt + i\int_0^1 (4t^3 - 2t^5)dt$$
$$= \left[\tfrac{1}{3}t^3 - t^5\right]_0^1 + i\left[t^4 - \tfrac{1}{3}t^6\right]_0^1$$
$$= \tfrac{2}{3}(i - 1).$$

EXAMPLE 2. Integrate $1/z$ around the unit circle C given by $z(t) = \cos t + i \sin t$ $(0 \leqslant t \leqslant 2\pi)$.

Since $z'(t) = -\sin t + i \cos t = i(\cos t + i \sin t)$, we have

$$\int_C 1/z\,dz = \int_0^{2\pi} (\cos t + i \sin t)^{-1}\,i(\cos t + i \sin t)\,dt$$

$$= i \int_0^{2\pi} dt$$

$$= 2\pi i.$$

If γ^* is the opposite path to γ, we have

$$\int_{\gamma^*} f(z) \, dz = \int_\alpha^\beta f(z(\alpha+\beta-t)) \frac{d}{dt} (z(\alpha+\beta-t)) \, dt.$$

Put $\alpha+\beta-t = s$, then the integral becomes

$$\int_\beta^\alpha f(z(s)) \frac{d}{ds} (z(s)) \frac{ds}{dt} \frac{dt}{ds} \, ds$$

$$= \int_\beta^\alpha f(z(s)) \frac{d}{ds} (z(s)) \, ds$$

$$= - \int_\alpha^\beta f(z(s)) \frac{d}{ds} (z(s)) \, ds$$

i.e. $$\int_{\gamma^*} f(z) \, dz = - \int_\gamma f(z) \, dz. \qquad (4)$$

Now suppose γ is a contour. Then γ consists of a finite number of smooth paths $\gamma_1, \ldots, \gamma_n$ and we define

$$\int_\gamma f(z) \, dz = \int_{\gamma_1} f(z) \, dz + \ldots + \int_{\gamma_n} f(z) \, dz \qquad (5)$$

EXAMPLE 3. Integrate z^2 along the contour γ given by

$$z(t) = \begin{cases} t & (0 \leqslant t \leqslant 1) \\ 1+i(t-1) & (1 \leqslant t \leqslant 2) \end{cases}$$

$$\int_\gamma z^2 dz = \int_0^1 t^2 . 1 dt + \int_1^2 (1+i(t-1))^2 . i dt$$

$$= \left[\tfrac{1}{3} t^3 \right]_0^1 + \int_1^2 (2-2t) \, dt + i \int_1^2 (2t-t^2) \, dt$$

$$= \tfrac{1}{3} + \left[2t-t^2 \right]_1^2 + i \left[t^2 - \tfrac{1}{3} t^3 \right]_1^2$$

$$= \tfrac{2}{3} (i-1).$$

Using (4), (5) we have the following rules for contour integration:

RULE 1. If γ is composed of two contours γ_1, γ_2,

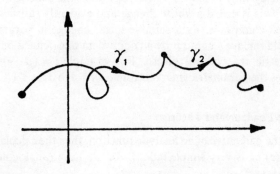

Figure 12

then $\int_\gamma f(z)\,dz = \int_{\gamma_1} f(z)\,dz + \int_{\gamma_2} f(z)\,dz$.

RULE 2. If γ^* is the opposite contour to γ, then

$$\int_{\gamma^*} f(z)\,dz = -\int_\gamma f(z)\,dz.$$

The value of a contour integral is unchanged when we change the parameter $t = h(u)$ $(a \leqslant u \leqslant b)$, where $h(a) = \alpha$, $h(b) = \beta$, h is (strictly) monotonic increasing and $h'(u)$ is continuous for $a \leqslant u \leqslant b$. We then have

$$\int_\gamma f(z)\,dz = \int_\alpha^\beta f(z)\,\frac{dz}{dt}\,dt = \int_a^b f(z)\,\frac{dz}{dt}\,\frac{dt}{du}\,du = \int_a^b f(z)\,\frac{dz}{du}\,du.$$

This is analogous to the real case†, and may be used to simplify the calculation.

We could consider two contours to be equivalent for the

† W. Ledermann, *Integral Calculus*, p. 12

purposes of integration if they are related to each other by a change in parameter as above. Two equivalent contours have the same track and are traversed in the same direction as the parameter increases. A particularly simple change in parameter is given by $h(u) = mu + c$ where m, c are real constants and $m > 0$. This is called a *linear* change and evidently satisfies the required conditions. By a suitable linear change in parameter we could replace a contour by an equivalent one defined on any parametric interval we please. For example $t = (\beta - \alpha)u + \alpha$ changes the parametric interval from $\alpha \leqslant t \leqslant \beta$ to $0 \leqslant u \leqslant 1$.

3. The Fundamental Theorem

If f is the derivative of an analytic function, then the calculation of $\int_\gamma f(z)\, dz$ is very simple indeed, for we have (analogous to the real case):

THE FUNDAMENTAL THEOREM OF CONTOUR INTEGRATION. Suppose that f is a continuous function defined in the domain D. If f is the derivative of an analytic function F in D, and γ is a contour in D starting at z_1 and ending at z_2, then

$$\int_\gamma f(z)\, dz = F(z_2) - F(z_1).$$

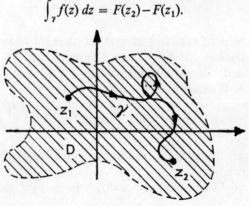

Figure 13

Proof. Write $f(z) = u(x, y) + iv(x, y)$, $F(z) = U(x, y) + iV(x, y)$ and suppose that γ is given by $z(t) = x(t) + iy(t)$ $(\alpha \leqslant t \leqslant \beta)$, where $z(\alpha) = z_1$, $z(\beta) = z_2$.

Since $f = F'$, using the Cauchy-Riemann equations for F, we have

$$F' = u + iv = \frac{\partial U}{\partial x} + i\frac{\partial V}{\partial x} = \frac{\partial V}{\partial y} - i\frac{\partial U}{\partial y}$$

and so $u = \dfrac{\partial U}{\partial x} = \dfrac{\partial V}{\partial y}, \qquad v = \dfrac{\partial V}{\partial x} = -\dfrac{\partial U}{\partial y}.$

Thus

$$\int_\gamma f(z)\, dz = \int_\alpha^\beta (u + iv)(x' + iy')\, dt$$

$$= \int_\alpha^\beta \left(\frac{\partial U}{\partial x} x' + \frac{\partial U}{\partial y} y'\right) dt + i\int_\alpha^\beta \left(\frac{\partial V}{\partial x} x' + \frac{\partial V}{\partial y} y'\right) dt.$$

But we have† $\dfrac{dU}{dt} = \dfrac{\partial U}{\partial x}\dfrac{dx}{dt} + \dfrac{\partial U}{\partial y}\dfrac{dy}{dt}$

and so $\displaystyle\int_\gamma f(z)\, dz = \int_\alpha^\beta \frac{dU}{dt}\, dt + i\int_\alpha^\beta \frac{dV}{dt}\, dt$

$$= F(z(\beta)) - F(z(\alpha))$$

$$= F(z_2) - F(z_1), \text{ as required.}$$

If $f = F'$ in D, then F is called a *primitive* of f in D. A primitive is unique up to an additive constant, because if $f = F_1', f = F_2'$ in D, then $F_1' - F_2' = 0$ and so $F_1 - F_2$ is constant in D.

Obviously the simplest way to integrate is to look for a primitive. For example $z^2 = \dfrac{d}{dz}\left(\tfrac{1}{3}z^3\right)$ and $\tfrac{1}{3}z^3$ is a primitive for z^2. Thus if γ is any contour joining 0 to $1+i$, we have

† P. Hilton, *Partial Derivatives*, pp. 12, 13.

$$\int_\gamma z^2 \, dz = \tfrac{1}{3}(1+i)^3 = \tfrac{2}{3}(1-i).$$

This integral has already been calculated for certain contours (examples 1, 3).

More generally, if n is an integer, $n \neq -1$, then $z^n = \dfrac{d}{dz}\left(\dfrac{z^{n+1}}{n+1}\right)$. This holds for all $n \geq 0$ and for $z \neq 0$ if $n \leq -2$. Hence if γ is a contour not passing through the origin which starts at z_1 and ends at z_2, then

$$\int_\gamma z^n \, dz = \frac{z_2^{n+1}}{n+1} - \frac{z_1^{n+1}}{n+1} \qquad (n \neq -1).$$

In particular, if γ is a closed contour, we have $z_1 = z_2$ and

$$\int_\gamma z^n \, dz = 0 \qquad (n \neq -1).$$

This illustrates the following consequences of the fundamental theorem:

COROLLARY 3.1. If f is the continuous derivative of an analytic function, then for any contour γ in the domain of definition of f, $\int_\gamma f(z) \, dz$ depends only on the endpoints and not on the particular contour.

COROLLARY 3.2. If f is the continuous derivative of an analytic function, then for any *closed* contour γ in the domain of definition, $\int_\gamma f(z) = 0$.

Proof of 3.2. $\int_\gamma f(z) \, dz = F(z_2) - F(z_1) = 0$, since $z_1 = z_2$.

Warning. Not every continuous function has a primitive. For such functions the fundamental theorem does not apply and the only way to evaluate the integral is by direct calculation from the basic definition. For such functions the integral *does* depend on the path, and the integral round a closed curve may not be zero.

If F is analytic in a domain D and $f = F'$, we will prove later (page 56) that f is also analytic. This remarkable theorem shows that if f is not analytic, then it cannot have a primitive F.

Even if f is analytic in D, it need not have a primitive defined throughout the whole of D. For example $f(z) = 1/z$ in the domain D consisting of all points except the origin. If f had a primitive in D, then $\int_\gamma 1/z\, dz = 0$ for any closed contour γ in D. But for the unit circle C given by $z(t) = \cos t + i \sin t \,(0 \leqslant t \leqslant 2\pi)$, we know (example 2) that $\int_C 1/z\, dz = 2\pi i \neq 0$. This shows that no primitive can exist.

We recall that $1/z = \dfrac{d}{dz}(\text{Log } z)$ in the cut-plane, but this does not lead to a contradiction since the unit circle crosses the negative real axis and so does not lie completely in the cut-plane.

Notice the striking difference between z^n (where n is an integer, $n \neq -1$) and z^{-1}, in that

$$\int_C z^n\, dz = \begin{cases} 2\pi i \text{ if } n = -1, \\ 0 \text{ otherwise.} \end{cases}$$

This result is responsible for much of the theory in Chapters II, III of Functions of a Complex Variable II.

We conclude this section with an inequality which will prove useful later:

THEOREM 3.3. *If γ is a contour of length L and $|f(z)| \leqslant M$ on the track of γ, then*

$$\left| \int_\gamma f(z)\, dz \right| \leqslant ML.$$

Proof. If γ is given by $z(t) = x(t) + iy(t)$ $(\alpha \leqslant t \leqslant \beta)$, then we recall† that the length of γ is given by

$$L = \int_\alpha^\beta \{(x'(t))^2 + (y'(t))^2\}^{\frac{1}{2}}\, dt = \int_\alpha^\beta |z'(t)|\, dt.$$

† W. Ledermann, *Multiple Integrals*, p. 4.

Assuming the inequality

$$\left| \int_{\alpha}^{\beta} g(t) \, dt \right| \leqslant \int_{\alpha}^{\beta} |g(t)| \, dt \tag{6}$$

for a complex function g of a real variable t, then

$$\left| \int_{\gamma} f(z) \, dz \right| = \left| \int_{\alpha}^{\beta} f(z(t)) \, z'(t) \, dt \right|$$

$$\leqslant \int_{\alpha}^{\beta} |f(z(t)) \, z'(t)| \, dt$$

$$\leqslant \int_{\alpha}^{\beta} M |z'(t)| \, dt$$

$$= ML$$

To verify (6), we use a simple trick.

Let $\int_{\alpha}^{\beta} g(t) \, dt = Re^{i\Theta}$ where R, Θ are real, $R \geqslant 0$. Then

$$R = \int_{\alpha}^{\beta} e^{-i\Theta} g(t) dt = \int_{\alpha}^{\beta} U(t) dt + i \int_{\alpha}^{\beta} V(t) \, dt$$

where

$$e^{-i\Theta} g(t) = U(t) + i V(t).$$

Since R is real, $\int_{\alpha}^{\beta} V(t) \, dt = 0$. But now, by the real case†, since $U(t) \leqslant |g(t)|$ we have

$$R = \int_{\alpha}^{\beta} U(t) \, dt \leqslant \int_{\alpha}^{\beta} |g(t)| \, dt, \text{ as required.}$$

REMARK. An upper bound M for $|f(z)|$ on the track of γ can always be found. We have not developed the technique for a neat proof, but we sketch an outline of a proof as follows:

The function $m(t) = |f(z(t))|$ is a continuous real-valued function of t for $\alpha \leqslant t \leqslant \beta$. By continuity at α, $m(t)$ must be bounded for $\alpha \leqslant t \leqslant \alpha + \varepsilon$ where $\varepsilon > 0$. Let x_0 be the upper bound of all points x in $\alpha \leqslant x \leqslant \beta$ such that $m(t)$ is bounded in $\alpha \leqslant t \leqslant x$. By continuity at x_0, $m(t)$ must be bounded in $\alpha \leqslant t \leqslant x_0$. By hypothesis we have $x_0 \leqslant \beta$. We cannot have $x_0 < \beta$ because the continuity of $m(t)$ would imply that $m(t)$ is

† W. Ledermann, *Integral Calculus*, p. 6.

bounded in a neighbourhood of x_0, i.e. for $|t-x| \leqslant \delta$ where $\delta > 0$. Thus we would have $m(t)$ bounded for $\alpha \leqslant t \leqslant x_0 + \delta$ (by the larger of the bounds in $\alpha \leqslant t \leqslant x_0$, $x_0 \leqslant t \leqslant x_0 + \delta$). This contradicts the definition of x_0. Hence $x_0 = \beta$ and since $m(t)$ is bounded for $\alpha \leqslant t \leqslant x_0$, we have the desired result.

4. Cauchy's Theorem

In the last section we were discussing conditions under which $\int_\gamma f(z)\,dz = 0$ for a closed contour γ. If f is assumed analytic in a domain containing γ. this need not be true. For example we have seen that $\int_C 1/z\,dz \neq 0$ where C is the unit circle $z(t) = \cos t + i \sin t \ (0 \leqslant t \leqslant 2\pi)$. The significant factor here is that $1/z$ is not analytic everywhere *inside* C. (It is not defined at the origin.)

CAUCHY'S THEOREM. If f is analytic in a domain D and γ is a closed Jordan contour in D whose interior also lies in D, then

$$\int_\gamma f(z)\,dz = 0.$$

It is not possible in this text to give a complete proof of this

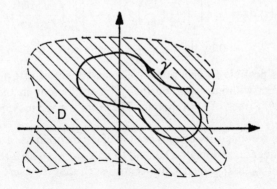

Figure 14

very deep result, but an outline will be given at the end of this section. In the original proof, Cauchy himself needed to assume that not only was f analytic (i.e. f' exists throughout D) but that f' was continuous. He gave several proofs, one of which used the Cauchy-Riemann equations.

$$\int_\gamma f(z)dz = \int_\gamma (u+iv)\,(dx+i\,dy)$$

$$= \int_\gamma (u\,dx - v\,dy) + i\int_\gamma (v\,dx + u\,dy).$$

Let A be the set of points inside and on the track of γ. Then Green's Theorem† states that under suitable conditions, including the continuity of $P(x, y)$, $Q(x, y)$, $\dfrac{\partial Q}{\partial x}$, $\dfrac{\partial P}{\partial y}$, we have

$$\int_\gamma (P\,dx + Q\,dy) = \iint_A \left(\frac{\partial Q}{\partial x} - \frac{\partial P}{\partial y}\right) dx\,dy.$$

If we assume f' is continuous, this implies the continuity of u, v, $\dfrac{\partial u}{\partial x}$, $\dfrac{\partial u}{\partial y}$, $\dfrac{\partial v}{\partial x}$, $\dfrac{\partial v}{\partial y}$. Hence

$$\int_\gamma f(z)\,dz = -\iint_A \left(\frac{\partial v}{\partial x} + \frac{\partial u}{\partial y}\right) dx\,dy + i\iint_A \left(\frac{\partial u}{\partial x} - \frac{\partial v}{\partial y}\right) dx\,dy$$

$$= 0 \text{ by the Cauchy-Riemann equations.}$$

It is possible to give a fairly elementary proof of Cauchy's Theorem without assuming that f' is continuous in the case where the track of γ is a triangle. The proof is given in Appendix II.

This seemingly innocuous version of the theorem has a strong consequence. A domain D is said to be a *star-domain* if there is a point z_0 in D (called a star-centre) such that for every other

† W. Ledermann, *Multiple Integrals*, p. 38.

point z in D, the whole straight line segment joining z_0 to z lies in D. Examples of star-domains are drawn in figure 15:

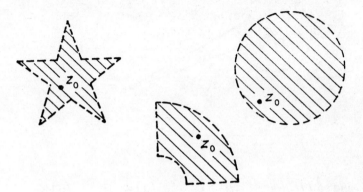

Figure 15

Inside a star-domain an analytic function always has a primitive:

PROPOSITION 4.1. If f is an analytic function defined in a star-domain D, then we may construct an analytic function F defined in D such that $F' = f$.

Proof. Denote by $[z_1, z_2]$ the contour $z(t) = z_1(1-t)+z_2 t$ $(0 \leqslant t \leqslant 1)$ which describes the straight line joining z_1 to z_2. If z_0 is the star-centre of D and z_1 is in D, then $[z_0, z_1]$ lies in D and we define

$$F(z_1) = \int_{[z_0, z_1]} f(z)\, dz$$

We will prove $F' = f$ and so F is a primitive for f.

Since D is open, there is an $\varepsilon_1 > 0$ such that for $|h| < \varepsilon_1$ we have $z_1 + h$ is in D, and evidently the line $[z_1, z_1 + h]$ lies in D. By Cauchy's Theorem for a Triangle (Appendix II), we have

$$\int_{[z_0, z_1]} f(z)\, dz + \int_{[z_1,\, z_1 + h]} f(z)\, dz + \int_{[z_1 + h,\, z_0]} f(z)\, dz = 0.$$

47

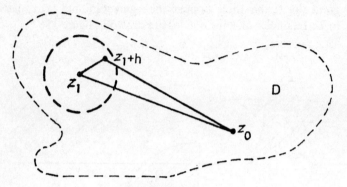

Figure 16

and so $F(z_1+h)-F(z_1) = \int_{[z_0, z_1+h]} f(z)\,dz - \int_{[z_0, z_1]} f(z)\,dz$

$$= \int_{[z_1, z_1+h]} f(z)\,dz.$$

Keeping z_1 constant, we have $\int_{[z_1, z_1+h]} f(z_1)\,dz = f(z_1)h$ and this gives

$$\frac{F(z_1+h)-F(z_1)}{h} - f(z_1) = \int_{[z_1, z_1+h]} \frac{\{f(z)-f(z_1)\}}{h}\,dz \quad (h\neq 0).$$

Since f is analytic in D, it is certainly continuous at z_1 and so given $\varepsilon > 0$, we have $|f(z)-f(z_1)| < \varepsilon$ for z in a neighbourhood of z_1. Also the length of $[z_1, z_1+h]$ is $|h|$ and so for sufficiently small h we have

$$\left| \frac{F(z_1+h)-F(z_1)}{h} - f(z_1) \right| \leqslant \frac{\varepsilon}{|h|} \cdot |h| = \varepsilon.$$

Since ε is arbitrary, this implies

$$\lim_{h\to 0} \frac{F(z_1+h)-F(z_1)}{h} = f(z_1)$$

i.e. $F'(z_1) = f(z_1)$.

48

Since z_1 is an arbitrary point in D, this completes the proof.

As immediate consequences of this proposition and corollaries 3.1, 3.2 of the fundamental theorem of contour integration, we have:

COROLLARY 4.2. If f is an analytic function defined in a star-domain D, then for any contour γ in D, $\int_\gamma f(z)\, dz$ depends only on the end-points of γ.

COROLLARY 4.3. If f is analytic in a star-domain D and γ is any closed contour in D, then $\int_\gamma f(z)\, dz = 0$.

We may use corollary 4.2 to sketch a proof of Cauchy's Theorem. First note that an open disc given by $|z - z_0| < r$ is a star-domain. If γ is a contour in an arbitrary domain D, it is always possible to subdivide γ into a finite number of sub-contours $\gamma_1, \ldots, \gamma_n$ where each γ_r lies in an open disc D_r which itself is contained in D. (The proof is omitted.) Let z_{r-1}, z_r be the initial and final points of γ_r.

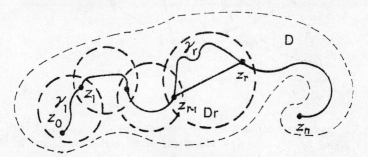

Figure 17

Since D_r is a star-domain, by corollary 4.2,

$$\int_{\gamma_r} f(z)\, dz = \int_{[z_{r-1}, z_r]} f(z)\, dz.$$

If P is the polygon with sides $[z_0, z_1], [z_1, z_2], \ldots, [z_{n-1}, z_n]$, then

$$\int_\gamma f(z)\,dz = \sum_{r=1}^n \int_{\gamma_r} f(z)\,dz = \int_P f(z)\,dz.$$

Now suppose γ is a *closed* Jordan contour in D whose interior also lies in D. To show $\int_\gamma f(z)\,dz = 0$, it is sufficient to prove $\int_P f(z)\,dz = 0$ for the closed polygon P. To do this, we may draw in extra lines joining vertices of P, making triangular contours $\Delta_1, \ldots, \Delta_m$ such that

(i) The track and interior of each Δ_r lies in D,

(ii) $\int_P f(z)\,dz = \sum_{r=1}^m \int_{\Delta_r} f(z)\,dz.$

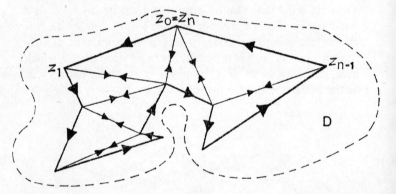

Figure 18

In any particular case this is geometrically obvious and as in figure 18, the integrals along the additional lines cancel in opposite pairs. Note, however, that it is difficult to write down a general rule as to how this is done. Assuming its validity, by (i) and Cauchy's Theorem for a triangle we have $\int_{\Delta_r} f(z)\,dz = 0$ and by (ii) we have $\int_P f(z)\,dz = 0$. Hence $\int_\gamma f(z)\,dz = 0$.

This concludes our discussion on Cauchy's Theorem.

EXERCISES ON CHAPTER TWO

In exercises 1–8 integrate the given function along the contour $z(t) = 1 - t + it^2 \ (0 \leqslant t \leqslant 1)$. (Use the Fundamental Theorem of Contour Integration wherever possible.)

1. $\mathscr{R}z$ 2. $1/z^3$ 3. $(4z^3 + z^4)e^z$ 4. $1/z$ 5. z^2 6. $\bar{z}$

7. $\sin^2 z$ 8. $z^\alpha = e^{\alpha \operatorname{Log} z} \ (\alpha \neq -1)$.

In exercises 9–11, integrate the given function around the unit circle $z(t) = \cos t + i \sin t \ (0 \leqslant t \leqslant 2\pi)$.

9. $1/z^2$ 10. $|z|$ 11. $\bar{z}$.

12. If $f(z) = c_0 + c_1 z + \ldots + c_n z^n + \ldots$ for $|z| < R$, prove $F(z)$
$= c_0 z + \dfrac{c_1 z^2}{2} + \ldots + \dfrac{c_n z^{n+1}}{n+1} + \ldots$ is absolutely convergent for $|z| < R$.

Use the result of Appendix I to show $F'(z) = f(z)$ for $|z| < R$. If γ is a contour in $|z| < R$, starting at the origin and finishing at z_0, show

$$\int_\gamma f(z)\,dz = c_0 z_0 + \frac{c_1 z_0^2}{2} + \ldots + \frac{c_n z_0^{n+1}}{n+1} + \ldots$$

(This states a power series may be integrated term by term inside the circle of convergence.)

CHAPTER THREE

Taylor's Series

---

1. Cauchy's Integral Formula

THEOREM 1.1. Suppose that f is an analytic function defined in a domain D. Let γ be a closed Jordan contour in D whose interior lies completely in D. If γ is described anti-clockwise (as the parameter increases) and z_0 is a point inside γ, then

$$f(z_0) = \frac{1}{2\pi i} \int_\gamma \frac{f(z)}{z - z_0}\, dz.$$

This is Cauchy's Integral Formula.

Proof. Let C_ε be the circle centre z_0, radius ε in the standard parametrization $z(t) = z_0 + \varepsilon e^{it}$ ($0 \leqslant t \leqslant 2\pi$), where ε is small so that the track of C_ε is inside that of γ.

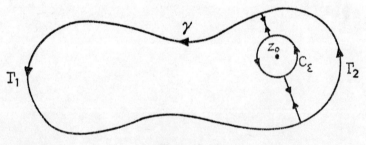

Figure 19

Make two cross-cuts from the track of C_ε to that of γ and parametrize them, making two Jordan contours Γ_1, Γ_2 where Γ_1, Γ_2 each traverse part of γ anti-clockwise, across a cut, round part of C_ε clockwise (i.e. the opposite sense to C_ε) and across the other cut as in figure 19. (We are relying on geometric intuition for this construction.)

The function $F(z) = \dfrac{f(z)}{z - z_0}$ is analytic inside and on Γ_r for

$r = 1, 2$ and by Cauchy's Theorem

$$\int_{\Gamma_r} F(z)\, dz = 0 \qquad r = 1, 2.$$

Adding these two integrals, the contributions due to the cross-cuts cancel and we have

$$\int_\gamma F(z)\, dz - \int_{C_\varepsilon} F(z)\, dz = 0. \qquad (1)$$

Now

$$\int_{C_\varepsilon} F(z)\, dz = \int_{C_\varepsilon} \frac{f(z_0)}{z - z_0}\, dz + \int_{C_\varepsilon} \frac{f(z) - f(z_0)}{z - z_0}\, dz. \qquad (2)$$

Since $\lim\limits_{z \to z_0} \dfrac{f(z) - f(z_0)}{z - z_0} = f'(z_0)$, for z near z_0 we must have

$\left| \dfrac{f(z) - f(z_0)}{z - z_0} \right| \leqslant M$ and so $\left| \displaystyle\int_{C_\varepsilon} \dfrac{f(z) - f(z_0)}{z - z_0}\, dz \right| \leqslant M.2\pi\varepsilon$

using theorem 3.3 of Chapter II. As $\varepsilon \to 0$, the contribution of this integral tends to zero.

Also

$$\int_{C_\varepsilon} \frac{f(z_0)}{z - z_0}\, dz = f(z_0) \int_0^{2\pi} \frac{1}{e^{it}}\, i e^{it}\, dt = f(z_0).2\pi i.$$

Substituting into (1), and letting $\varepsilon \to 0$, we find

$$\int_\gamma \frac{f(z)}{z - z_0}\, dz = f(z_0).2\pi i,$$

which completes the proof.

Note that this remarkable theorem shows that the values of an analytic function at all points inside a closed Jordan contour are uniquely determined by the values of the function on that contour.

2. Taylor's Series

In this section we use Cauchy's integral formula to express an analytic function as a power series in a neighbourhood of a point.

LEMMA 2.1. If f is analytic in the open disc given by $|z-z_0| < R$, then $f(z_0+h) = a_0+a_1h+ \ldots +a_nh^n+ \ldots$ for $|h| < R$. If C is a circle centre z_0, radius r where $0 < r < R$, given by $z(t) = z_0+re^{it}$ $(0 \leqslant t \leqslant 2\pi)$ then $a_n = \dfrac{1}{2\pi i} \displaystyle\int_C \dfrac{f(z)}{(z-z_0)^{n+1}}dz$.

Proof. Fix h where $|h| < R$ and initially restrict r to $|h| < r < R$. By Cauchy's integral formula

$$f(z_0+h) = \frac{1}{2\pi i} \int_C \frac{f(z)}{z-z_0-h} \, dz.$$

But $1/(z-z_0-h) = \dfrac{1}{z-z_0} \left(1-\dfrac{h}{z-z_0}\right)^{-1} = \dfrac{1}{z-z_0} \; (1-w)^{-1}$

where $w = \dfrac{h}{z-z_0}$. Since $1+w+ \ldots +w^{n-1} = \dfrac{1-w^n}{1-w}$, we have

$$(1-w)^{-1} = 1+w+ \ldots +w^{n-1}+\frac{w^n}{1-w}.$$

Substituting into the integral formula for $f(z_0+h)$ and simplifying, we obtain

$$f(z_0+h) = \frac{1}{2\pi i} \int_C f(z)\left\{\frac{1}{z-z_0}+\frac{h}{(z-z_0)^2}+ \ldots +\frac{h^{n-1}}{(z-z_0)^n}\right.$$

$$\left.+\frac{h^n}{(z-z_0)^n(z-z_0-h)}\right\} dz = a_0+a_1h+ \ldots +a_{n-1}h^{n-1}+A_n$$

where

$$a_m = \frac{1}{2\pi i} \int_C \frac{f(z)}{(z-z_0)^{m+1}} \, dz$$

and

$$A_n = \frac{1}{2\pi i} \int_C \frac{f(z)h^n}{(z-z_0)^n(z-z_0-h)} \, dz.$$

For z on the track of C we have $|f(z)| \leqslant M$ where M is some real constant. Moreover for z on the track of C, $|z-z_0| = r$ and $|z-z_0-h| \geqslant ||z-z_0| - |h|| = r - |h|$, hence

$$|A_n| \leqslant \frac{1}{2\pi} \frac{M|h|^n}{r^n(r-|h|)} 2\pi r = \frac{Mr}{(r-|h|)} \left(\frac{|h|}{r}\right)^n.$$

Since $|h| < r$, we have $A_n \to 0$ as $n \to \infty$ and so the infinite series $\sum a_n h^n$ converges to the sum $f(z_0 + h)$.

Note that we have only proved the expression

$$a_n = \frac{1}{2\pi i} \int_C \frac{f(z)}{(z-z_0)^{n+1}} \, dz$$

for $|h| < r < R$, but since the integral is independent of h, no matter how small, the expression must be true for any r satisfying $0 < r < R$.

If we write $z = z_0 + h$ then we have

$$f(z) = a_0 + a_1(z-z_0) + \ldots + a_n(z-z_0)^n + \ldots \text{ for } |z-z_0| < R.$$

But a power series is differentiable as many times as we please inside its circle of convergence (proposition 6.1, Chapter I) and

$$a_n = \frac{f^{(n)}(z_0)}{n!}.$$

Now suppose that f is defined on an arbitrary domain D. If z_0 is in D then, by definition, so is an ε-neighbourhood given by $|z-z_0| < \varepsilon$. If R is the largest such ε (possible infinite), then using lemma 2.1 inside $|z-z_0| < R$ we obtain:

TAYLOR'S THEOREM. If f is analytic in a domain D, then

f is differentiable as many times as we please throughout D. If z_0 is in D, then

$$f(z) = f(z_0) + f'(z_0)(z - z_0) + \ldots +$$
$$+ \frac{f^{(n)}(z_0)}{n!}(z - z_0)^n + \ldots \qquad |z - z_0| < R,$$

where $|z - z_0| < R$ is the largest open disc centre z_0 contained in D.

Substituting $z = z_0 + h$, the power series may also be written as:

$$f(z_0 + h) = f(z_0) + f'(z_0)h + \ldots + \frac{f^{(n)}(z_0)}{n!} h^n + \ldots \qquad |h| < R.$$

REMARK. The importance of this phenomenal result cannot be over-emphasized. We need only assume a complex function is differentiable once throughout its domain of definition and then it is infinitely differentiable. This contrasts strongly with the real case, where a function may be differentiable once but not twice (refer back to page 30 for an example).

EXAMPLE 1. $f(z) = \text{Log } z$ is analytic in the cut-plane. The largest open disc centre $z_0 = 1$ in the cut-plane is $|z - 1| < 1$. Since $f^{(n)}(z_0) = \dfrac{(-1)^n(n-1)!}{z_0^n} = (-1)^n(n-1)!$, we have:

$$\text{Log}(1 + h) = h - \frac{h^2}{2} + \ldots + \frac{(-1)^n h^n}{n} + \ldots \qquad |h| < 1.$$

EXAMPLE 2. $f(z) = 1/z$ is analytic for $z \neq 0$. If $z_0 \neq 0$, then

$$1/(z_0 + h) = \frac{1}{z_0}\left(1 + \frac{h}{z_0}\right)^{-1}$$
$$= \frac{1}{z_0} - \frac{h}{z_0^2} + \frac{h^2}{z_0^3} - \ldots + \frac{(-1)^n h^n}{z_0^{n+1}} + \ldots \qquad \left|\frac{h}{z_0}\right| < 1.$$

If $\left|\dfrac{h}{z_0}\right| < 1$, then $|h| < |z_0|$, which states that $z_0 + h$ lies in the circle centre z_0, radius $|z_0|$. This is the largest circle centre z_0 which does not include the origin (where $1/z$ is not defined). Note that the coefficient of h^n in the power series is

$$\frac{(-1)^n}{z_0^{n+1}} = \frac{1}{n!} f^{(n)}(z_0).$$

In the notation of lemma 2.1,

$$f^{(n)}(z_0) = n!a_n = \frac{n!}{2\pi i} \int_C \frac{f(z)}{(z-z_0)^{n+1}}\, dz,$$

where C is a circle centre z_0, lying in the domain of definition of f. This is Cauchy's Formula for the nth derivative of f.

If $|f(z)| \leqslant M$ on the circle C centre z_0, radius r, then

$$|f^{(n)}(z_0)| = \frac{n!}{2\pi} \left| \int_C \frac{f(z)}{(z-z_0)^{n+1}}\, dz \right|$$

$$\leqslant \frac{n!}{2\pi} \frac{M}{r^{n+1}} . 2\pi r$$

and so

$$|f^{(n)}(z_0)| \leqslant \frac{Mn!}{r^n}.$$

This result is called *Cauchy's Inequality*. Using it we may prove:

LIOUVILLE'S THEOREM. If f is analytic throughout the whole plane and $|f(z)| \leqslant M$ for all z, then f is a constant function.

Proof. $|f'(z_0)| \leqslant \dfrac{M}{r}$ for *any* r, since f is analytic throughout the whole plane. Let $r \to \infty$ and we have $f'(z_0) = 0$. This is true for any z_0 and so $f'(z) = 0$ for all z. This implies that f is constant.

3. Zeros and the Identity Theorem

If $f(z_0) = 0$, we say that z_0 is a *zero of f*. We may write an analytic function f as a power series in a neighbourhood of z_0,

$$f(z) = a_0 + a_1(z-z_0) + \ldots + a_n(z-z_0)^n + \ldots \quad |z-z_0| < R.$$

Either $a_n = 0$ for all n, in which case $f(z) = 0$ for $|z-z_0| < R$, or we have $a_0 = a_1 = \ldots = a_{m-1} = 0$ and $a_m \neq 0$. In the latter case we say that z_0 is a *zero of order m*. Note that since $a_n = f^{(n)}(z_0)/n!$, a zero of order m satisfies $f(z_0) = 0, f'(z_0) = 0, \ldots, f^{(m-1)}(z_0) = 0$, but $f^{(m)}(z_0) \neq 0$.

We may show that a zero of order m is isolated. By this we mean that there is an ε-neighbourhood of z_0 in which z_0 is the *only* zero of f, i.e. $f(z) \neq 0$ for $0 < |z-z_0| < \varepsilon$.

To see this we write

$$f(z) = (z-z_0)^m \{a_m + a_{m+1}(z-z_0) + \ldots\} \text{ for } |z-z_0| < R$$
$$= (z-z_0)^m \Phi(z)$$

where the power series $\Phi(z) = a_m + a_{m+1}(z-z_0) + \ldots$ is convergent for $|z-z_0| < R$. Since Φ is analytic for $|z-z_0| < R$, it is certainly continuous at z_0 and so $\Phi(z) \to a_m \neq 0$ as $z \to z_0$. Hence $\Phi(z)$ is non-zero in some ε-neighbourhood of z_0. But $(z-z_0)^m$ is zero only at z_0 and so $f(z) \neq 0$ for $0 < |z-z_0| < \varepsilon$.

Suppose that $z_1, z_2, \ldots, z_n, \ldots$ is a sequence of distinct† zeros of f which tends to a point z_0. If f is defined at z_0, then by continuity we must have $f(z_0) = 0$. Since z_0 is a limit of zeros, it is not isolated, and as we have seen above we must have f identically zero inside some circle centre z_0.

Note. This argument depends on the fact that f is analytic in a neighbourhood of z_0, in particular it breaks down if f is

† We consider the zeros to be distinct to avoid the trivial case that all but a finite number of the zeros coincide at z_0.

not defined or not analytic at z_0. For example the function f given by $f(z) = \sin(1/z)$ $(z \neq 0)$ is analytic everywhere except at the origin. It has a sequence of zeros given by $z_n = 1/n\pi$ $(n \geqslant 1)$ which tends to the origin, but evidently the function is not identically zero inside any circle with the origin as centre.

THEOREM 3.1. Suppose that $z_1, z_2, \ldots, z_n, \ldots$ is a sequence of distinct zeros of an analytic function f defined in a domain D and that the limit of this sequence, z_0, lies in D, then f is identically zero throughout D.

Proof. By continuity $f(z_0) = 0$ and, as we have seen above, f is identically zero inside some circle $|z - z_0| < \varepsilon_0$.

Let w be any other point in D. Since D is a domain, there is a stepwise curve in D joining z_0 to w. We suppose that this curve has length d and let $z(s)$ be the point distance s along it from z_0 so that $z(0) = z_0$ and $z(d) = w$. We intend to show that $f(z) = 0$ all along the curve, in particular $f(w) = 0$.

Since f is identically zero in $|z - z_0| < \varepsilon_0$, then $f(z(s))$ is certainly zero for $0 \leqslant s < \varepsilon_0$. We consider those real numbers s in $0 \leqslant s \leqslant d$ such that $f(z) = 0$ along the curve as far as $z(s)$. Suppose that s^* is the least upper bound of such s. By continuity $f(z(s^*)) = 0$ and $z(s^*)$ is the furthest point along the curve such that $f(z) = 0$ for all z on the curve as far as $z(s^*)$ (marked with a thick line in figure 20).

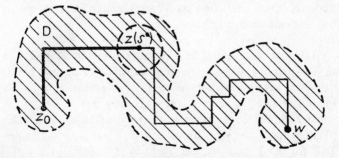

Figure 20

We cannot have $z(s^*) \neq w$. This is because $f(z) = 0$ along the curve up to $z(s^*)$ and so $f(z) = 0$ in a neighbourhood $|z - z(s^*)| < \varepsilon$. This would imply that $f(z) = 0$ for a certain distance along the curve beyond $z(s^*)$, contradicting the definition of s^*. Hence $z(s^*) = w$ and $f(w) = 0$. This completes the proof.

We may immediately deduce:

THE IDENTITY THEOREM. Suppose that f, g are analytic functions defined in the same domain D. Let $z_1, z_2, \ldots, z_n, \ldots$ be a sequence of distinct points in D with limit z_0 also in D, such that $f(z_n) = g(z_n)$ for $n \geqslant 1$, then $f(z) = g(z)$ throughout D.

Proof. Apply theorem 3.1 to $\Phi(z) = f(z) - g(z)$, then Φ is analytic in D and $z_1, z_2, \ldots, z_n, \ldots$ is a sequence of distinct zeros of Φ with limit z_0 in D.

The Identity Theorem has far reaching consequences in the theory of analytic functions.

Suppose that f_0 is a complex function defined on a set S. A complex function f is said to be an *extension* of f_0 if f is defined on a larger set D containing S and $f(z) = f_0(z)$ for all z in S. In general the values of f at points outside S can be assigned quite arbitrarily. For example if S is the real axis and $f(x) = \sin x$ for x real, then defining $f(z) = \sin z$ for z on the real axis and $f(z) = 0$ otherwise, the function f is an extension of f_0. However if we insist that the extension is also *analytic*, then (under a minor restriction on S) the Identity Theorem shows that this extension is *unique*.

THEOREM 3.2. Let f_0 be a complex function defined on a set S which contains a convergent sequence of distinct points $z_1, z_2, \ldots$ together with its limit. If f is an extension of f_0 to a domain D and f is analytic, then f is unique.

Proof. Suppose that g is another analytic function defined in D satisfying $g(z) = f_0(z)$ for all z in S. Then $g(z_n) = f_0(z_n) = f(z_n)$ for $n \geqslant 1$ and by the Identity Theorem, $g(z) = f(z)$ throughout D.

We remark that the notion of extension to an analytic function does not guarantee that such an extension exists, only that if it exists then its uniqueness is assured. It also does not give any practical method of constructing an extension and we will find that usually the most successful way is to resort to inspired guesswork.

EXAMPLE. $f_0(z) = 1 + z + z^2 + \ldots + z^n + \ldots \qquad |z| < 1$.
The set S of complex numbers satisfying $|z| < 1$ certainly contains a convergent sequence of points $\left(\text{e.g. } \dfrac{1}{2}, \dfrac{1}{3}, \ldots, \dfrac{1}{n+1}, \right.$
$\ldots$ with limit $0 \Big)$ and so an extension to an analytic function in any given domain D is unique. The power series for f_0 is not convergent for $|z| > 1$ but the function $f(z) = (1-z)^{-1}$ is analytic for $z \neq 1$ and satisfies $f(z) = f_0(z)$ for $|z| < 1$. Hence the analytic function f is the extension of f_0 to the domain consisting of all complex numbers except $z = 1$. Note however that no analytic function exists which is an extension to the whole plane, because $f(z)$ has no finite limit as $z \to 1$ and so we cannot define $f(1)$ in any way to make f analytic there.

The notion of extension by an analytic function is particularly interesting in two cases:

CASE I. S is the real axis (or more generally any subset of the real axis containing a convergent sequence of distinct points together with its limit). Given a real-valued function f_0 defined on S, if there is an extension to a complex analytic function f in some domain containing S then this function is unique. This shows the strong restriction imposed on a complex function by requiring it to be analytic.

For example if $f_0(x) = \sin x$ for all real x, then of course we know that $f(z) = \sin z$ gives an analytic function defined throughout the whole plane and f coincides with f_0 on the real axis. We now know that $f(z) = \sin z$ is the *only* analytic function which satisfies $f(x) = \sin x$ for x real.

CASE II. S is a non-empty open set.

Since S is non-empty it contains a point z_0 and since it is open it includes an ε-neighbourhood of z_0. We can easily select a sequence of distinct points in this neighbourhood which tends to z_0 $\left(\text{e.g. } z_0 + \frac{1}{2}\varepsilon, \ z_0 + \frac{1}{3}\varepsilon, \ \ldots, \ z_0 + \frac{1}{n+1}\varepsilon, \ \ldots \right)$ and so S satisfies the required conditions. As a particular instance we may take S to be the open disc $|z - z_0| < \varepsilon$.

We have seen in section 2 that if f is an analytic function defined in a domain D and z_0 is in D, then f has a Taylor series expansion in a small disc centre z_0. The notion of extension using an analytic function shows that the reverse process is true in that once we know the values in a small disc in D then the values of f throughout D are uniquely determined. Hence in some peculiar way the power series expansion in a small disc contains all the information required to specify the values of the function throughout its domain of definition!

EXERCISES ON CHAPTER THREE

Find the Taylor expansion of the following analytic functions at the origin:

1. $z(1-z)^{-2}$ 2. $z^3 e^z$ 3. $(z+1)^3$ 4. $\text{Log}(1+z)$ 5. $(1+z)^\alpha$ 6. $(1+z^2)^{-1}$.

7. Suppose that f is analytic throughout the whole plane and satisfies $|f(z)| \leqslant M|z|^n$ for all z. Use Cauchy's inequality to prove that $f^{(n+1)}(z) = 0$ and show that $f(z)$ is polynomial of degree at most n.

Find the extension of the power series in 8–10 to analytic functions in the largest possible domain.

8. $\displaystyle\sum_{n=1}^{\infty} nz^n \quad |z| < 1$ 9. $\displaystyle\sum_{n=1}^{\infty} n^2 z^n \quad |z| < 1$ (Hint: differentiate $\sum nz^n$)

10. $\displaystyle\sum_{n=0}^{\infty} (-1)^n z^{2n} \quad |z| < 1$.

11. If $a_n(z) = z^n + (1-z)^n$, by considering $\displaystyle\sum_{n=0}^{N} a_n(z)$, prove that $\displaystyle\sum_{n=0}^{\infty} a_n(z)$ converges if $|z| < 1$ and $|1-z| < 1$. Draw the domain given by $|z| < 1$ and $|1-z| < 1$. Find the sum $f(z) = \displaystyle\sum_{n=0}^{\infty} a_n(z)$ in this domain and hence write down the extension of $f(z)$ to an analytic function in the largest possible domain.

12. Suppose that f is analytic in a domain D containing the point

$$z = 1 \text{ and } f\left(1 - \frac{1}{k}\right) = \sum_{n=0}^{\infty} (-1)^n \left(1 - \frac{1}{k}\right)^{2n} \text{ for } k = 1, 2, \ldots .$$

Calculate the following (if they exist):

$$f(0), f(1+i), f(i), f(2,000).$$

Appendix I

THEOREM. If $f(z) = c_0 + c_1 z + c_2 z^2 + \ldots + c_n z^n + \ldots$ for $|z| < R$, then $f'(z) = c_1 + 2c_2 z + \ldots + nc_n z^{n-1} + \ldots$ for $|z| < R$.

Proof. First we show that the power series

$$f_1(z) = c_1 + 2c_2 z + \ldots + nc_n z^{n-1} + \ldots$$

is absolutely convergent for $|z| < R$.

Fix z and choose r such that $|z| < r < R$.

By hypothesis $\sum_{n=0}^{\infty} c_n r^n$ converges absolutely and so there is some positive number K such that $|c_n r^n| < K$ for all n.

Let $q = \dfrac{|z|}{r}$, then $0 \leqslant q < 1$ and $|nc_n z^{n-1}| \leqslant \dfrac{nK|z|^{n-1}}{r^n} = \dfrac{Knq^{n-1}}{r}$.

But $\sum_{n=0}^{\infty} nq^{n-1}$ converges (to $(1-q)^{-2}$), hence by the comparison test, $\sum_{n=0}^{\infty} nc_n z^{n-1}$ converges absolutely.

Now we show $f'(z_0) = f_1(z_0)$ for $|z_0| < R$, i.e.

$$\lim_{z \to z_0} \left\{ \frac{f(z) - f(z_0)}{z - z_0} - f_1(z_0) \right\} = 0.$$

As before, choose r such that $|z_0| < r < R$ and since $z \to z_0$, we may also restrict z so that $|z| < r$.

We know $\sum_{n=0}^{\infty} nc_n r^{n-1}$ converges absolutely. Suppose we are given $\varepsilon > 0$, then we can find an integer N such that $\sum_{n=N}^{\infty} |nc_n r^{n-1}| < \frac{1}{4}\varepsilon$. Now keep N fixed. We can write

64

$$\frac{f(z)-f(z_0)}{z-z_0}-f_1(z) =$$

$$\sum_{n=0}^{\infty} c_n \{z^{n-1}+z_0 z^{n-2}+ \ldots +z_0^{n-1}-nz_0^{n-1}\}$$

We let $\sum_1$ be the sum of the first N terms of this series (i.e. from $n = 0$ to $n = N-1$) and $\sum_2$ the sum of the remaining terms. Then

$$\left|\sum_2\right| \leqslant \sum_{n=N}^{\infty} |c_n| \{r^{n-1}+r^{n-1}+ \ldots +r^{n-1}+nr^{n-1}\}$$

$$= \sum_{n=N}^{\infty} 2n|c_n|r^{n-1} < \tfrac{1}{2}\varepsilon.$$

Also $\sum_1 = \sum_{n=0}^{N} c_n \{z^{n-1}+z_0 z^{n-2}+ \ldots +z_0^{n-1}-nz_0^{n-1}\}$ is a polynomial in z and $\lim_{z \to z_0} \sum_1 = 0$. Hence there is a $\delta > 0$ such that $\left|\sum_1\right| < \tfrac{1}{2}\varepsilon$ provided that $|z-z_0| < \delta$. Thus for $|z| < r$ and $|z-z_0| < \delta$ we have

$$\left|\frac{f(z)-f(z_0)}{z-z_0}-f_1(z_0)\right| \leqslant \left|\sum_1\right|+\left|\sum_2\right| < \tfrac{1}{2}\varepsilon+\tfrac{1}{2}\varepsilon = \varepsilon.$$

This means $f'(z_0) = f_1(z_0)$ as required.

Appendix II

CAUCHY'S THEOREM FOR A TRIANGLE

Suppose f is analytic in a domain D and T is a triangular contour whose track and interior lie in D, then $\int_T f(z)\, dz = 0$.

Proof. Suppose $|\int_T f(z)\, dz| = h \geqslant 0$, then by a neat trick we show $h = 0$.

Draw in lines joining the midpoints of the sides of T and parametrize them, giving four triangular contours $T^{(1)}$, $T^{(2)}$, $T^{(3)}$, $T^{(4)}$, such that integrals along the additional lines cancel in pairs because they are taken in opposite directions.

If $I_n = \int_{T^{(n)}} f(z)\, dz$, $n = 1, 2, 3, 4$,
then

$$I_1 + I_2 + I_3 + I_4 = \int_T f(z)\, dz.$$

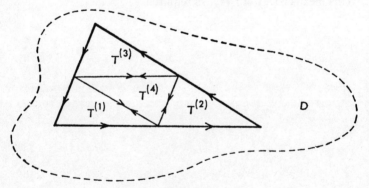

Figure 21

Since $|\int_T f(z)\,dz| = h$, we can choose r such that $|I_r| \geqslant \tfrac{1}{4}h$. Define $T_1 = T^{(r)}$, then

$$\left|\int_{T_1} f(z)\,dz\right| \geqslant \tfrac{1}{4}h$$

and since T_1 is half the linear size of T, the perimeter length of T_1 is given by

$$L(T_1) = \tfrac{1}{2}L(T).$$

Repeat the process of subdivision with T_1 and so on, obtaining a sequence of triangles $T_1, T_2, \ldots, T_n, \ldots$ where

$$\left|\int_{T_n} f(z)\,dz\right| \geqslant (\tfrac{1}{4})^n h \tag{1}$$

and

$$L(T_n) = (\tfrac{1}{2})^n L(T) \tag{2}$$

This sequence of triangles approaches some point z_0 which lies inside or on the triangle T. By hypothesis, f is analytic at z_0 and so

$$\lim_{z \to z_0} \left\{ \frac{f(z) - f(z_0)}{z - z_0} \right\} = f'(z_0).$$

This means that given any $\varepsilon > 0$, we can find $\delta > 0$ such that if $|z - z_0| < \delta$, then

$$\left| \frac{f(z) - f(z_0)}{z - z_0} - f'(z_0) \right| < \varepsilon. \tag{3}$$

The condition $|z - z_0| < \delta$ means z lies in a disc centre z_0, radius δ. Since the sequence of triangles $T_1, T_2, \ldots, T_n, \ldots$ approaches z_0 and each T_n is half the linear dimensions of its predecessor T_{n-1}, for some N we have T_n lying inside this disc for $n \geqslant N$. Thus for all z on the triangle T_n, $n \geqslant N$, from (3) we have

$$|f(z) - f(z_0) - f'(z_0)\,(z - z_0)| \leqslant \varepsilon|z - z_0| \leqslant \varepsilon L(T_n). \tag{4}$$

By the Fundamental Theorem of Contour Integration, since T_n is a closed contour, $\displaystyle\int_{T_n} dz = \int_{T_n} \frac{d}{dz}(z)\,dz = 0$ and $\displaystyle\int_{T_n} z\,dz$

$$= \int_{T_n} \frac{d}{dz}\left(\frac{1}{2}z^2\right) dz = 0.$$ Since z_0 is fixed, we see that

$$\int_{T_n} f(z)\, dz = \int_{T_n} \{f(z) - f(z_0) - f'(z_0)\,(z - z_0)\}\, dz.$$

From (4), we have

$$\left| \int_{T_n} f(z)\, dz \right| \leqslant \varepsilon L(T_n).L(T_n)$$
$$= \varepsilon(\tfrac{1}{4})^n L(T)^2 \qquad \text{from (2)}$$

Comparing this with (1), we find

$$(\tfrac{1}{4})^n h \leqslant \varepsilon(\tfrac{1}{4})^n L(T)^2$$

i.e. $$h \leqslant \varepsilon L(T)^2.$$

But $h \geqslant 0$ and ε may be arbitrarily small. This implies that $h = 0$.

Solutions to Exercises

Chapter One

1. (i) $x^2 - y^2 + 2x + i(2xy + 2y)$ (ii) $x(x^2 + y^2)^{-1} - iy(x^2 + y^2)^{-1}$
(iii) $\sin x \cosh y + i \cos x \sinh y$
(iv) $(xe^x \cos y - x + ye^x \sin y)(e^{2x} - 2e^x \cos y + 1)^{-1}$
$\qquad + i(ye^x \cos y - y - xe^x \sin y)(e^{2x} - 2e^x \cos y + 1)^{-1}$
(v) $\frac{1}{2} \log (x^2 + y^2) + i \tan^{-1}(y/x)$ where we choose $0 \leqslant \tan^{-1}(y/x) < \pi$ for $y \geqslant 0$ and $-\pi < \tan^{-1}(y/x) \leqslant 0$ for $y \leqslant 0$.
(vi) $x^2 + y^2 + i.0$ (vii) $\tan^{-1}(y/x)$ with the conventions of 1.(v) above.

2. (i) $2z + 2$ (ii) $-z^{-2}$ (iii) $\cos z$ (iv) $(e^z - 1 - ze^z)(e^z - 1)^{-2}$
(v) z^{-1}.

3. (a) $\dfrac{\partial u}{\partial x} = 2x, \dfrac{\partial u}{\partial y} = 2y, \dfrac{\partial v}{\partial x} = 0 = \dfrac{\partial v}{\partial y}$.

At $z = 0$, $\dfrac{d}{dz}(|z|^2) = \lim_{z \to 0} \dfrac{|z|^2 - |0|^2}{z} = \lim_{z \to 0} \dfrac{z\bar{z}}{z} = 0$.

(b) $\dfrac{\partial u}{\partial x} = \dfrac{-y}{x^2 + y^2}$ $\dfrac{\partial u}{\partial y} = \dfrac{x}{x^2 + y^2}$, $\dfrac{\partial v}{\partial x} = 0 = \dfrac{\partial v}{\partial y}$.

4. (i) domain (ii) no (not open) (iii) no (not connected) (iv) yes
(v) yes.

5. $f = u + iv$ where $v = 0$. $\dfrac{\partial v}{\partial x} = \dfrac{\partial v}{\partial y} = 0$. By the Cauchy-Riemann equations $\dfrac{\partial u}{\partial x} = \dfrac{\partial u}{\partial y} = 0$. Thus $f' = 0$ throughout the domain of definition which implies that f is constant.

6. (i) $\lambda_1 = ik, \lambda_2 = -ik$ (ii) $Ae^z + Be^{2z}$ (iii) $Ae^z + Be^{2iz} + Ce^{-2iz}$.

7. $(1 - z)^{-4} = 1 + 4z + \dfrac{4.5}{1.2} z^2 + \ldots + \dfrac{(n+1)(n+2)(n+3)}{6} z^n + \ldots$

69

Chapter Two

(In questions marked * the integration may be performed using the Fundamental Theorem of Contour Integration).

1. $-\dfrac{1}{2}+\dfrac{1}{3}i$ 2*. 1 3*. $e^{i}-e$ 4*. $\text{Log } i - \text{Log } 1 = i\dfrac{\pi}{2}$ 5*. $-\dfrac{1}{3}i-\dfrac{1}{3}$

6. $\dfrac{2}{3}i$ 7*. $\dfrac{1}{2}i-\dfrac{1}{4}i\sinh 2 - \dfrac{1}{2}+\dfrac{1}{4}\sin 2$

8*. $\dfrac{i^{\alpha+1}}{\alpha+1} - \dfrac{1^{\alpha+1}}{\alpha+1} = \dfrac{e^{i(\pi/2)(\alpha+1)}-1}{\alpha+1}$ (since $i^{\alpha+1} = e^{(\alpha+1)\text{Log }i} = e^{i(\pi/2)(\alpha+1)}$)

9*. 0 10. 0 11. $2\pi i$.

12. (i) $\displaystyle\sum \frac{c_n z^{n+1}}{n}$ converges absolutely by comparison with $\sum |c_n z^n|$ since $\left|\dfrac{c_n z^{n+1}}{n}\right| \Big/ |c_n z^n| = \dfrac{|z|}{n} \to 0$ as $n \to \infty$.

(ii) $\int_\gamma f(z)\,dz = F(z_0) - F(0) = \displaystyle\sum \frac{c^n z_0^{n+1}}{n}$.

Chapter Three

1. $z+2z^2+ \ldots +nz^n+ \ldots$ $|z|<1$ 2. $z^3+z^4+ \ldots +\dfrac{z^{n+3}}{n!}+ \ldots$ for all z.

3. $1+3z+3z^2+z^3$ for all z 4. $z-\dfrac{z^2}{2} \ldots +(-1)^n\dfrac{z^{n+1}}{n+1}+ \ldots$ $|z|<1$.

5. $1+\alpha z+ \ldots +\dfrac{\alpha(\alpha-1) \ldots (\alpha-n+1)z^n}{n!}+ \ldots$ $|z|<1$ (unless α is a positive integer in which case the series terminates and is valid for all z).

6. $1-z^2+ \ldots +(-1)^n z^{2n}+ \ldots$ $|z|<1$.

7. If $|z - z_0| = r$ and $r \geq |z_0|$, then $|z| \leq |z_0| + |z - z_0| \leq 2r$ and so $|f(z)| \leq 2^n r^n M$. Hence $|f^{(n+1)}(z_0)| \leq \dfrac{2^n r^n M (n+1)!}{r^{n+1}}$. Let $r \to \infty$, then $f^{(n+1)}(z_0) = 0$.

8. $\dfrac{z}{(1-z)^2} \qquad (z \neq 1)$

   9. $\dfrac{z(1+z)}{(1-z)^3} \qquad (z \neq 1)$

10. $(1+z^2)^{-1} \qquad (z \neq \pm i)$.

11. $f(z) = \dfrac{z}{1-z} + \dfrac{1}{z} \qquad (z \neq 0, 1)$.

12. $f(z) = \displaystyle\sum_{n=0}^{\infty} (-1)^n z^{2n} = (1+z^2)^{-1}$ wherever f is defined. Hence $f(0) = 1$, $f(1+i) = (1-2i)/5$, $f(2{,}000) = 1/4{,}000{,}001$ (if they are defined) but $f(i)$ cannot exist-if f is analytic.

FUNCTIONS
OF A COMPLEX
VARIABLE
PART II

Preface to Part II

This part follows on from Part I and uses the ideas of complex differentiation and integration developed there.

First there is a geometrical interpretation of analytic functions in terms of conformal mappings which is then shown to have a substantial link with harmonic functions and two-dimensional potential theory. The next two chapters are concerned with the development of the Calculus of Residues by way of Laurent series and Cauchy's Residue Theorem. Many worked examples of calculation of integrals by residues are given together with the method of calculating the sum of certain series.

The final chapter is an account of analytic continuation and Riemann surfaces. This is given in descriptive terms (often in terms of examples) in the hope of explaining these two ideas and the way that they complement each other.

CHAPTER ONE

Conformal Mappings and Harmonic Functions

1. Conformal Mappings

In this section we discuss the geometrical properties of analytic functions. First we calculate the gradient of a smooth path in the complex plane.

If $z_0 \neq z_1$, then $\theta = \arg(z_1 - z_0)$ $(-\pi < \theta \leqslant \pi)$ is the angle between the real axis and the directed line from z_0 to z_1. Suppose that $z(t) = x(t) + iy(t)$ $(\alpha \leqslant t \leqslant \beta)$ is a smooth path† and $z_0 = z(t_0)$, $z_1 = z(t)$ are two distinct points on its track, then $\theta = \arg(z(t) - z(t_0))$ is the angle between the real axis and the directed chord from $z(t_0)$ to $z(t)$.

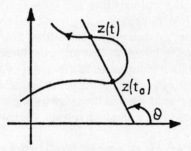

Figure 1

Now if c is a positive real number, then $\arg cz = \arg z$. If we assume that $t > t_0$ then $\dfrac{1}{t - t_0} > 0$ and so

$$\theta = \arg(z(t) - z(t_0)) = \arg\left\{\frac{z(t) - z(t_0)}{t - t_0}\right\}.$$

† i.e. $z'(t) = x'(t) + iy'(t)$ exists and is continuous for $\alpha \leqslant t \leqslant \beta$.

9

Let $t \to t_0$, then the chord tends to the tangent at t_0 directed in the sense t increasing. Also $\dfrac{z(t) - z(t_0)}{t - t_0} \to z'(t_0)$. From this we may infer that the angle between the real axis and the directed tangent is arg $z'(t_0)$, provided that $z'(t_0) \neq 0$.

The case $z'(t_0) = 0$ is omitted because arg 0 is not well-defined. The proof in other cases is not trivial because arg z denotes the *principal* value $-\pi < \arg z \leqslant \pi$, and arg is not continuous on the negative real axis. Let $w = \dfrac{z(t) - z(t_0)}{t - t_0}$, $w_0 = z'(t_0)$. Since arg is continuous in the cut-plane†, when $-\pi < \arg w_0 < \pi$ we have $w \to w_0$ implies arg $w \to$ arg w_0. Thus $\theta \to$ arg $z'(t_0)$. However if arg $w_0 = \pi$, i.e. if w_0 is on the negative real axis, then although arg $w_0 = \pi$, a point near w_0 but below the real axis has arg w nearly $-\pi$. If w tends to w_0 from below the real axis then arg $w_0 \to -\pi$. Worse still, if w tends to w_0 in a spiral path, going round and round and getting ever closer to w_0 then arg w jumps from nearly $-\pi$ to π and back again ad infinitum so that arg w does not tend to a limit. Thus it is blatantly untrue to say that $w \to w_0$ implies arg $w \to$ arg w_0 in the case of the principal value. If arg $w_0 = \pi$, we choose the value of arg w in the range $0 < \arg w \leqslant 2\pi$. This value is continuous near w_0 and as $w \to w_0$, we have arg $w \to \pi$, as required.

Now suppose f is an analytic function defined on a domain D. Let γ be a smooth path in D given by $z(t) = x(t) + iy(t)$ $(\alpha \leqslant t \leqslant \beta)$, then f transforms γ into a smooth path Γ given by $w(t) = f(z(t))$ $(\alpha \leqslant t \leqslant \beta)$. Suppose that z_0 is a point in D where $f'(z_0) \neq 0$ and z_0 lies on the track of γ, i.e. $z_0 = z(t_0)$ for some t_0.

We compare the directions of the tangent to γ at z_0 and the

† *Functions of a Complex Variable I*, p. 18.

tangent to Γ at $w_0 = f(z_0)$. Let $\phi = \arg z'(t_0)$, $\psi = \arg w'(t_0)$. Since

$$w'(t_0) = f'(z(t_0))z'(t_0)$$

we have $\arg w'(t_0) = \arg f'(z(t_0)) + \arg z'(t_0)$ up to a multiple of 2π and so $\psi = \arg f'(z_0) + \phi$ up to a multiple of 2π.

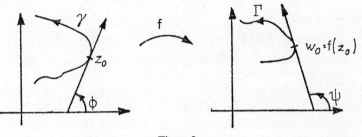

Figure 2

Hence the tangent to γ at z_0 is turned through an angle $\arg f'(z_0)$ upon transformation under f. This does not depend on the path γ and so if γ_1, γ_2 are two paths through z_0, then the transformed paths meet at the same angle† as γ_1, γ_2. (In each case the tangent is turned through the same angle $\arg f'(z_0)$, up to a multiple of 2π, upon transformation.)

A transformation preserving angles between curves is said to be *conformal*. An analytic function is conformal where $f'(z) \neq 0$. (It is certainly not conformal where $f'(z) = 0$. If z_0 is a zero of order m of f', then the angle between curves through z_0 is multiplied by $m+1$ upon transformation. The proof is omitted.)

We can find more information about analytic functions by considering the equation

$$\lim_{z \to z_0} \frac{f(z) - f(z_0)}{z - z_0} = f'(z_0).$$

† The angle between two paths through z_0 is the angle between their tangents (considered up to a multiple of 2π).

11

This implies

$$\lim_{z \to z_0} \left| \frac{f(z) - f(z_0)}{z - z_0} \right| = |f'(z_0)|$$

and so for z near z_0, we have

$$\left| \frac{f(z) - f(z_0)}{z - z_0} \right| \simeq |f'(z_0)|$$

i.e. $|f(z) - f(z_0)| \simeq |f'(z_0)| \, |z - z_0|$.

This says that f magnifies lengths by approximately $|f'(z_0)|$ near z_0.

Taking z_0, z_1, z_2 'close together', where $f'(z_0) \neq 0$, then conformality and the magnification property state that the small triangle with vertices z_0, z_1, z_2 is transformed into a similar triangle, with sidelengths multiplied approximately by $|f'(z_0)|$ and turned through an angle $\arg f'(z_0)$. The smaller the triangle, the better the approximation.

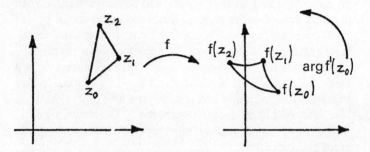

Figure 3

As an example of a conformal mapping†, we consider $f(z) = \dfrac{az + b}{cz + d}$ $(ad \neq bc)$ which is defined for all z if $c = 0$ and

† 'Mapping' is just another word for 'function'.

12

for all z except $z = -d/c$ otherwise. This is called a *bilinear mapping*. Note that $f'(z) = \dfrac{ad-bc}{(cz+d)^2}$ and so the condition $ad \neq bc$ ensures that $f'(z) \neq 0$ wherever f is defined and so f is conformal.

As particular cases we note:

EXAMPLE 1. A *translation* $w = z + \alpha$. Points in the w-plane correspond to those in the z-plane with a change in origin. Figures remain the same shape and size when transformed.

EXAMPLE 2. A *rotation* $w = e^{i\phi}z$ where ϕ is real. Since $\arg w = \arg z + \phi$ (up to a multiple of 2π) and $|w| = |z|$, we see that figures are rotated through an angle ϕ about the origin but lengths remain unchanged.

EXAMPLE 3. A *magnification* $w = rz$ where r is real and positive. A figure remains similar and similarly situated when transformed, but lengths are multiplied by a factor r.

EXAMPLE 4. An *inversion* $w = 1/z$. If $z = re^{i\theta}$ then $w = \dfrac{1}{r}e^{-i\theta}$ and so $|w| = 1/|z|$, $\arg w = -\arg z$. Unlike the previous examples, this may change the shape of figures. For example a circle may be transformed either into a circle or into a straight line. However, by considering a line to be a 'circle of infinite radius'†, it may be shown that an inversion transforms 'circles' into 'circles'. Other curves may have their shape altered, but because of the conformal property, the angle between two paths remains unaltered (provided that their intersection is not the origin, where the transformation is not defined).

† See Exercise 4 at the end of this chapter.

13

The reader is encouraged to draw pictures for the above examples to help visualize them.

It is a remarkable fact that a general bilinear mapping may be expressed as a succession of the particular types described above. For $c \neq 0$, we write

$$\frac{az+b}{cz+d} = \frac{bc-ad}{c^2(z+(d/c))} + \frac{a}{c}.$$

Let $\dfrac{bc-ad}{c^2} = \lambda$, then $\lambda \neq 0$. We write $w_1 = z+(d/c)$, $w_2 = 1/w_1$, $w_3 = |\lambda|w_2$, $w_4 = (\lambda/|\lambda|)w_3$, $w = w_4+(a/c)$. By successive substitution we find that w is obtained from z by a translation, then an inversion, a magnification, a rotation and another translation.

The case $c = 0$ is somewhat easier. We have $w = \dfrac{az+b}{d}$ $= \alpha z + \beta$ where $\alpha = a/d$, $\beta = b/d$. Thus if $w_1 = |\alpha|z$, $w_2 = (\alpha/|\alpha|)w_1$, $w = w_2+\beta$, we see that w is obtained from z by a magnification, a rotation and a translation.

Of the particular examples considered, only an inversion changes the shape of a figure and even this takes 'circles' into 'circles'. Thus a general bilinear mapping transforms 'circles' into 'circles'.

Bilinear mappings have many other interesting properties. The reader should consult the literature on the subject.†

2. Orthogonal Curves

As we have seen in the last section, the angle between two smooth paths is preserved under transformation by an analytic function where that function has non-zero derivative. The most important case occurs when the paths are orthogonal

† L. V. Ahlfors, *Complex Analysis*, McGraw Hill Book Co., pp. 76–88.

(i.e. intersect at right angles). If γ_1 is a line parallel to the x-axis and γ_2 is parallel to the y-axis, then they are orthogonal and so the transformed curves Γ_1, Γ_2 meet at right angles:

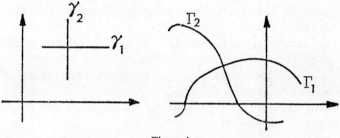

Figure 4

As an example of this phenomenon, consider the function $f(z) = e^z = e^{x+iy}$. Taking polar coordinates in the w-plane, $w = Re^{i\phi}$, then $w = f(z)$ gives $R = e^x$ and $\phi = y$ (up to a multiple of 2π). Thus the line $x = $ constant transforms into $R = $ constant, which is a circle centre the origin, and $y = $ constant transforms into $\phi = $ constant, which is a straight line through the origin. These evidently meet at right angles.

A most useful technique is to write $f(z) = u(x, y) + iv(x, y)$ and consider the curves $u(x, y) = u_0 = $ constant and $v(x, y) = v_0 = $ constant. Suppose that these are smooth paths

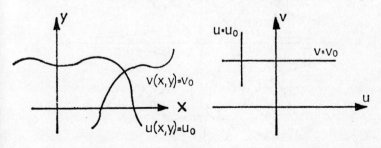

Figure 5

which meet in a point $z_0 = x_0 + iy_0$ where $f'(z_0) \neq 0$. If $w = u + iv = f(z)$, then the curve $u(x, y) = u_0$ in the z-plane transforms into $u = u_0$ in the w-plane and $v(x, y) = v_0$ transforms into $v = v_0$. But $u = u_0$, $v = v_0$ are straight lines parallel to the axes in the w-plane and hence meet at right angles.

This means that $u(x, y) = u_0$, $v(x, y) = v_0$ are orthogonal curves.

For different values of u_0, v_0 we obtain two families of curves. Any curve of the first family meets one of the second family at right angles. These curves are called the *level curves* of f.

EXAMPLE. $f(z) = z^2$. Since $f'(z) = 2z$, the transformation is conformal where $f'(z) \neq 0$, i.e. at all points except the origin. We have

$$f(z) = (x + iy)^2 = x^2 - y^2 + 2ixy$$

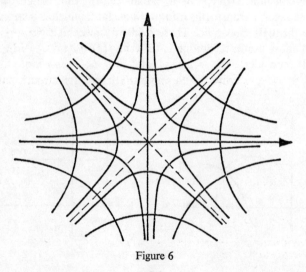

Figure 6

and so $u(x, y) = x^2 - y^2$, $v(x, y) = 2xy$. The level curves are $x^2 - y^2 = c$, $2xy = k$. For $c \neq 0$, $k \neq 0$, these do not pass through the origin and are hence orthogonal. Using coordinate geometry, for different values of c, the first set of curves are rectangular hyperbolae with asymptotes $x = y$, $x = -y$. Similarly the second set of curves are rectangular hyperbolae with the axes as asymptotes.

3. Harmonic Functions and Potential Theory

Suppose that $\phi(x, y)$ is a real valued function of two real variables x, y. If ϕ satisfies the differential equation

$$\frac{\partial^2 \phi}{\partial x^2} + \frac{\partial^2 \phi}{\partial y^2} = 0 \tag{1}$$

then ϕ is called a *harmonic function* or *potential function*. Equation (1) is called *Laplace's equation*.

Usually ϕ is only defined for those values of x, y where $x + iy$ lies in a domain D. If f is an analytic function defined in D and $f(z) = u(x, y) + iv(x, y)$, then it may be shown that both u and v are harmonic in D.

This follows from the Cauchy-Riemann equations† and Taylor's Theorem.‡ First note that

$$f'(z) = \frac{\partial u}{\partial x} + i\frac{\partial v}{\partial x} = \frac{\partial v}{\partial y} - i\frac{\partial u}{\partial y}. \tag{2}$$

From Taylor's Theorem, f'' exists throughout D and so f' is also analytic in D. Let $f' = U + iV$, then from the Cauchy-Riemann equations for U, V, the partial derivatives of U, V exist and satisfy:

---

† *Functions of a Complex Variable I*, p. 23.
‡ *Functions of a Complex Variable I*, p. 55.

$$\frac{\partial U}{\partial x} = \frac{\partial V}{\partial y} \tag{3}$$

$$\frac{\partial V}{\partial x} = -\frac{\partial U}{\partial y}. \tag{4}$$

Since $U = \dfrac{\partial u}{\partial x} = \dfrac{\partial v}{\partial y}$, $V = \dfrac{\partial v}{\partial x} = -\dfrac{\partial u}{\partial y}$, on substituting in (3) we have

$$\frac{\partial}{\partial x}\left(\frac{\partial u}{\partial x}\right) = \frac{\partial}{\partial x}\left(\frac{\partial v}{\partial y}\right) = \frac{\partial}{\partial y}\left(\frac{\partial v}{\partial x}\right) = -\frac{\partial}{\partial y}\left(\frac{\partial u}{\partial y}\right)$$

and in particular,

$$\frac{\partial^2 u}{\partial x^2} + \frac{\partial^2 u}{\partial y^2} = 0. \tag{5}$$

Substituting in (4), we also find

$$\frac{\partial}{\partial x}\left(\frac{\partial v}{\partial x}\right) = -\frac{\partial}{\partial x}\left(\frac{\partial u}{\partial y}\right) = -\frac{\partial}{\partial y}\left(\frac{\partial u}{\partial x}\right) = -\frac{\partial}{\partial y}\left(\frac{\partial v}{\partial y}\right)$$

which gives

$$\frac{\partial^2 v}{\partial x^2} + \frac{\partial^2 v}{\partial y^2} = 0. \tag{6}$$

This has applications in two-dimensional potential theory. If u is a potential function, then the curves $u(x, y) = $ constant are 'equipotential lines'. But we have seen in the last section that the curves $v(x, y) = $ constant are orthogonal to these. The curves $v(x, y) = $ constant are 'stream lines'.

Suppose that we are given a potential function u in a domain D. Is it possible to determine the equations of the stream lines from this? Under suitable conditions this problem may be solved by looking for a real-valued function v such that $f = u + iv$ is analytic in D. The function v is called the *harmonic conjugate* of u. If a harmonic conjugate exists, then from the Cauchy-Riemann equations for f we have

$$f' = \frac{\partial u}{\partial x} - i\frac{\partial u}{\partial y}. \tag{7}$$

From this we may attempt to find f. Sometimes the solution is obvious by inspection, otherwise we may use contour integration. The latter method would require restrictions on the nature of the domain D. For example, if D were a star-domain† with star-centre z_0, then we may adopt the method of Volume I, Chapter Two, proposition 4.1 to find

$$f(z_1) = \int_{[z_0, z_1]} f'(z)dz$$

$$= \int_{[z_0, z_1]} \left(\frac{\partial u}{\partial x} - i\frac{\partial u}{\partial y}\right)dz$$

where $[z_0, z_1]$ is the straight line from z_0 to z_1.

Note that a solution of (7) is unique up to an additive constant, for if f_1, f_2 are both solutions, then $f_1' = f_2'$. Hence $\frac{d}{dz}(f_1 - f_2) = 0$, and since D is a domain, $f_1 - f_2$ is constant throughout D (Volume I, Chapter One, theorem 5.1). This also implies that the harmonic conjugate v is unique up to an additive constant.

EXAMPLE 1. $u(x, y) = x^2 - y^2$, defined in the whole plane. Note first of all that u satisfies Laplace's equation. If f exists, then $f'(z) = \frac{\partial u}{\partial x} - i\frac{\partial u}{\partial y} = 2x - i(-2y) = 2z$, hence

$$f(z) = z^2 + \text{constant} = u + iv$$

and so $v(x, y) = 2xy + \text{constant}$.

† *Functions of a Complex Variable I*, p. 46.

EXAMPLE 2. $u(x, y) = \log\sqrt{(x^2+y^2)}$ defined in the whole plane except the origin. Since $\dfrac{\partial u}{\partial x} = \dfrac{x}{x^2+y^2}$, $\dfrac{\partial^2 u}{\partial x^2} = \dfrac{y^2-x^2}{(x^2+y^2)^2}$, $\dfrac{\partial u}{\partial y} = \dfrac{y}{x^2+y^2}$, $\dfrac{\partial^2 u}{\partial y^2} = \dfrac{x^2-y^2}{(x^2+y^2)^2}$, we see that u satisfies Laplace's equation throughout its domain of definition. If u were the real part of an analytic function f, then we would require $f'(z) = \dfrac{x}{x^2+y^2} - i\dfrac{y}{x^2+y^2} = \dfrac{1}{z}$. As we have shown (Volume I, page 43), no such f exists which is defined for all points except the origin. However, in the cut-plane (with the negative real axis removed) a solution is $f(z) = \text{Log } z$ and $v(x, y) = \arg(x+iy)$.

EXERCISES ON CHAPTER ONE

1. Consider the paths $z(t) = t$ $(-1 \leqslant t \leqslant 1)$, $z(t) = t(1+i)$ $(-1 \leqslant t \leqslant 1)$. Write down the equations of the transformed curves under the following functions: (i) e^z (ii) $\sin z$ (iii) $z^2 + z$. In each case verify that the function has non-zero derivative at the origin and that the angle between the curves is preserved under the transformation.

2. Consider the line segments $z(t) = t$ $(0 \leqslant t \leqslant 1)$, $z(t) = te^{i\alpha}$ $(0 \leqslant t \leqslant 1)$ where $-\pi < \alpha \leqslant \pi$. Find the equations of the transformed curves under the function $f(z) = z^n$ where n is a positive integer. Show that on transformation the angle between the two curves is multiplied by n (up to a multiple of 2π).

3. Find the equations of the level curves of $f(z) = \dfrac{1}{z}$ and draw a sketch of them.

4. Show that the equation of any circle or straight line may be written as

$$\varepsilon(x^2+y^2)+px+qy+r = 0 \qquad (*)$$

where p, q, r are real.

If $\varepsilon \neq 0$, show that this is a circle of radius $\left(\dfrac{p^2+q^2-4r\varepsilon}{4\varepsilon^2}\right)^{\frac{1}{2}}$

and if $\varepsilon = 0$ then it is a line. (This demonstrates why we regard a line as a 'circle of infinite radius'.)

Show that an inversion $w = 1/z$ transforms (*) into

$$r(u^2+v^2)+pu-qv+\varepsilon = 0$$

where $w = u+iv$.

Hence show that under an inversion
   (i) a straight line or circle through the origin transforms into a straight line,
   (ii) any other straight line or circle transforms into a circle.

5. Find the most general cubic form

$$u(x, y) = ax^3+bx^2y+cxy^2+dy^3 \quad (a, b, c, d \text{ real})$$

which satisfies Laplace's equation, and find an analytic function f which has u as its real part.

6. Verify that

$$u(x, y) = 2 \sin x \cosh y - 2\cos x \sinh y + x^2 - y^2 - 4xy$$

satisfies Laplace's equation, and (preferably by inspection) find an analytic function f which has u as its real part.

CHAPTER TWO

Cauchy's Residue Theorem

1. Laurent's Theorem

The main purpose of the next two chapters is to develop methods of calculating contour integrals. If f is analytic in a domain containing a closed Jordan contour γ and the points inside γ, then Cauchy's Theorem states that

$$\int_\gamma f(z)dz = 0.$$

In this chapter we are concerned with calculating $\int_\gamma f(z)dz$ where f is not analytic at a finite number of points inside γ. The solution to this problem is given by Cauchy's Residue Theorem which will then be used in Chapter Three to calculate a number of specific integrals.

We first generalize Taylor's Theorem. This states that if f is analytic for $|z-z_0| < R$, then we have a power series expansion $f(z_0+h) = \sum_{n=0}^{\infty} a_n h^n$, valid for $|h| < R$. Now suppose that f is only assumed analytic for $R_1 < |z-z_0| < R_2$. We cannot hope to express $f(z_0+h)$ as a power series $\sum_{n=0}^{\infty} a_n h^n$ valid for $R_1 < |h| < R_2$, since by the comparison test this series would converge for $|h| < R_2$, and by extending the domain of definition of $\sum a_n h^n$ to $|h| < R_2$ we may consider f to be analytic for $|z-z_0| < R_2$. We can however express $f(z_0+h)$ as a series involving both positive and negative powers of h.

LAURENT'S THEOREM. If f is analytic in the annulus $R_1 < |z-z_0| < R_2$ (where $R_1 \geqslant 0$), then

$$f(z_0+h) = \sum_{n=0}^{\infty} a_n h^n + \sum_{n=1}^{\infty} b_n h^{-n} \text{ for } R_1 < |h| < R_2.$$

If C is the circle centre z_0, radius r (given by $z(t) = z_0 + re^{it}$ ($0 \leqslant t \leqslant 2\pi$)) where $R_1 < r < R_2$, then

$$a_n = \frac{1}{2\pi i} \int_C \frac{f(z)}{(z-z_0)^{n+1}} \, dz, \, b_n = \frac{1}{2\pi i} \int_C (z-z_0)^{n-1} f(z) dz.$$

Note: If f is analytic for $R_1 < |z-z_0|$, we may formally take $R_2 = \infty$.

The proof is by expressing $f(z_0+h)$ in terms of two integrals; one is shown to equal $\sum_{n=0}^{\infty} a_n h^n$ for $|h| < R_2$ and the other $\sum_{n=1}^{\infty} b_n h^{-n}$ for $|h| > R_1$. Finding the two integrals is quite straightforward. To express each integral as a series is a little more technical, but is modelled on the proof of Taylor's Theorem (as in lemma 3.1, Chapter Three of Volume I). We now give the details.

Fix h and choose r_1, r_2 such that $R_1 < r_1 < |h| < r_2 < R_2$. Let C_m be the circular contour $z(t) = z_0 + r_m e^{it}$ ($0 \leqslant t \leqslant 2\pi$) for $m = 1, 2$. Note that $z_0 + h$ lies between C_1 and C_2. By making

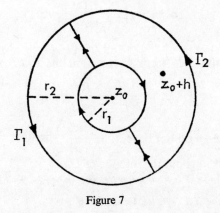

Figure 7

two cross-cuts from C_1 to C_2 avoiding the point $z_0 + h$, let Γ_1, Γ_2 be the two closed Jordan contours as in figure 7.

Then

$$\int_{\Gamma_1} \frac{f(z)}{z-(z_0+h)} \, dz = 0 \text{ by Cauchy's Theorem}$$

and

$$\int_{\Gamma_2} \frac{f(z)}{z-(z_0+h)} \, dz = 2\pi i f(z_0+h) \text{ by Cauchy's integral formula.}$$

Adding these integrals, the contributions along the cross-cuts cancel giving

$$f(z_0+h) = \frac{1}{2\pi i} \int_{C_2} \frac{f(z)}{z-z_0-h} \, dz - \frac{1}{2\pi i} \int_{C_1} \frac{f(z)}{z-z_0-h} \, dz.$$

As in the proof of Taylor's Theorem† we find that

$$\frac{1}{2\pi i} \int_{C_2} \frac{f(z)}{z-z_0-h} \, dz = \sum_{n=0}^{\infty} a_n h^n \qquad |h| < r_2$$

where

$$a_n = \frac{1}{2\pi i} \int_{C_2} \frac{f(z)}{(z-z_0)^{n+1}} \, dz.$$

But if C is any circle $z(t) = z_0 + re^{it}$ $(0 \leqslant t \leqslant 2\pi)$, $R_1 < r < R_2$, by making cross-cuts from C_2 to C in the usual way we find

$$a_n = \frac{1}{2\pi i} \int_C \frac{f(z)}{(z-z_0)^{n+1}} \, dz.$$

(Note that we do not have $a_n = \dfrac{f^{(n)}(z_0)}{n!}$ because f may not be analytic throughout the interior of C.)

Also, since $\sum a_n h^n$ converges for $|h| < r_2 < R_2$, by choosing

† *Functions of a Complex Variable I*, p. 54.

r_2 as close as we please to R_2, we find that $\sum a_n h^n$ converges for $|h| < R_2$.

Similarly

$$-\frac{1}{2\pi i}\int_{C_1} \frac{f(z)}{z-z_0-h}\, dz$$

$$= \frac{1}{2\pi i}\int_{C_1} f(z)\left\{\frac{1}{h}+\frac{z-z_0}{h^2}+\ldots+\frac{(z-z_0)^{n-1}}{h^n}-\frac{(z-z_0)^n}{h^n(z-z_0-h)}\right\}dz$$

$$= \sum_{r=1}^{n} b_r h^{-r} - B_n$$

where

$$b_r = \frac{1}{2\pi i}\int_{C_1} f(z)(z-z_0)^{n-1}dz, \quad B_n = \frac{1}{2\pi i}\int_{C_1} \frac{f(z)(z-z_0)^n}{h^n(z-z_0-h)}\, dz.$$

Now for some constant M we have $|f(z)| \leqslant M$ for z on the track of C_1. Moreover for such z we have $|z-z_0| = r_1$ and $|z-z_0-h| \geqslant ||h|-|z-z_0|| = |h|-r_1$. Hence

$$|B_n| \leqslant \frac{1}{2\pi}\frac{Mr_1^n}{|h|^n(|h|-r_1)}\cdot 2\pi r_1 = \frac{Mr_1}{|h|-r_1}\left(\frac{r_1}{|h|}\right)^n.$$

Since $r_1 < |h|$, we have $B_n \to 0$ as $n \to \infty$, and so

$$-\frac{1}{2\pi i}\int_{C_1} \frac{f(z)}{z-z_0-h}\, dz = \sum_{n=1}^{\infty} b_n h^{-n} \text{ for } |h| > r_1.$$

Arguing as for a_n, we find $b_n = \frac{1}{2\pi i}\int_{C} f(z)(z-z_0)^{n-1}dz$ where C is the circle $z(t) = z_0 + re^{it}$ ($0 \leqslant t \leqslant 2\pi$) for *any* r in $R_1 < r < R_2$. Also, by choosing r_1 as close to R_1 as we please, we find the series $\sum_{n=1}^{\infty} b_n h^{-n}$ converges for $|h| > R_1$.

This completes the proof.

Remark. By writing $b_n = a_{-n}$ for $n \geqslant 1$, we can express the result in a more symmetric form as

$$f(z_0+h) = \sum_{-\infty}^{\infty} a_n h^n \text{ where } a_n = \frac{1}{2\pi i}\int_C \frac{f(z)}{(z-z_0)^{n+1}} \, dz.$$

The integral formula allows us to show that the Laurent expansion is unique. That is to say that if we find $f(z) = \sum_{-\infty}^{\infty} c_n(z-z_0)^n$ by some other method, then $c_n = a_n$ for all n.

First note that $(z-z_0)^n = \dfrac{d}{dz}\left(\dfrac{(z-z_0)^{n+1}}{n+1}\right)$ $(n \neq -1)$, and so

$$\int_C (z-z_0)^n dz = 0 \qquad (n \neq -1).$$

If we recall that C is given by $z(t) = z_0 + re^{it}$ $(0 \leqslant t \leqslant 2\pi)$, then by direct calculation

$$\int_C \frac{1}{z-z_0} \, dz = \int_0^{2\pi} \frac{1}{re^{it}} \, ire^{it} \, dt = 2\pi i.$$

Hence, assuming term by term integration is justified in the annulus, we have

$$\int_C (z-z_0)^{-m-1}\left\{\sum_{-\infty}^{\infty} c_n(z-z_0)^n\right\}dz = \sum_{-\infty}^{\infty} c_n \int_C (z-z_0)^{n-m-1} dz$$
$$= 2\pi i c_m. \qquad (1)$$

This gives

$$a_m = \frac{1}{2\pi i}\int_C \frac{f(z)}{(z-z_0)^{m+1}} \, dz = c_m.$$

The integration in (1) is easily justified. We write

$$\sum_{-\infty}^{\infty} c_n(z-z_0)^{n-m-1} = \left(\ldots + \frac{c_{m-2}}{(z-z_0)^3} + \frac{c_{m-1}}{(z-z_0)^2}\right)$$
$$+ \frac{c_m}{z-z_0} + (c_{m+1} + c_{m+2}(z-z_0) + \ldots)$$
$$= f_1(z) + \frac{c_m}{z-z_0} + f_2(z).$$

We need only show that f_1, f_2 each have a primitive in the annulus and then

$$\int_C \left\{ \sum_{-\infty}^{\infty} c_n(z-z_0)^{n-m-1} \right\} dz = \int_C \left\{ f_1(z) + \frac{c_m}{z-z_0} + f_2(z) \right\} dz$$

$$= 0 + 2\pi i c_m + 0$$

$$= 2\pi i c_m \text{ as required.}$$

But
$$f_2(z) = \sum_{n=1}^{\infty} c_{m+n}(z-z_0)^{n-1} \qquad |z-z_0| < R_2$$

and so if
$$F_2(z) = \sum_{n=1}^{\infty} \frac{c_{m+n}}{n}(z-z_0)^n \qquad |z-z_0| < R_2$$

then $\dfrac{d}{dz}F_2(z) = f_2(z)$ and F_2 is a primitive for f_2.

For f_1, we have $f_1(z) = \sum_{n=1}^{\infty} \dfrac{c_{m-n}}{(z-z_0)^{n+1}} = \sum_{n=1}^{\infty} c_{m-n}w^{n+1}$ where $w = (z-z_0)^{-1}$. This is valid for $|z-z_0| > R_1$, i.e. for $|w| < 1/R_1$. If we choose

$$G(w) = -\sum_{n=1}^{\infty} \frac{c_{m-n}}{n} w^n \quad |w| < 1/R_1$$

then
$$\frac{d}{dw} G(w) = -\sum_{n=1}^{\infty} c_{m-n}w^{n-1}.$$

Hence if $F_1(z) = G((z-z_0)^{-1})$

$$= -\sum_{n=1}^{\infty} \frac{c_{m-n}}{n}(z-z_0)^{-n} \quad |z-z_0| > R_1,$$

then $\dfrac{d}{dz}F_1(z) = -(z-z_0)^{-2}G'((z-z_0)^{-1})$

$$= (z-z_0)^{-2} \sum_{n=1}^{\infty} c_{m-n}(z-z_0)^{-n+1} = f_1(z), \text{ as required.}$$

2. Isolated Singularities

If f is analytic in $0 < |z-z_0| < R$ we say that z_0 is an *isolated singularity* of f. For example, if $f(z) = 1/z$, then the origin is an isolated singularity. However if $f(z) = \text{Log } z$ in the cut-plane, then the origin is not an isolated singularity since every annulus $0 < |z| < R$ contains points on the negative real axis where $\text{Log } z$ is not analytic.

By Laurent's Theorem (with $R_1 = 0$ and $z = z_0 + h$), near an isolated singularity we may write

$$f(z) = \sum_{n=0}^{\infty} a_n(z-z_0)^n + \sum_{n=1}^{\infty} b_n(z-z_0)^{-n} \text{ for } 0 < |z-z_0| < R.$$

The series $\sum_{n=1}^{\infty} b_n(z-z_0)^{-n}$ is called the *principal part* of f at z_0. The behaviour of f near z_0 depends on the nature of the principal part and we distinguish three cases.

CASE 1.

The principal part is zero, i.e. every b_n is zero. Here z_0 is called a *removable singularity*, for we have

$$f(z) = \sum_{n=0}^{\infty} a_n(z-z_0)^n \qquad 0 < |z-z_0| < R$$

and by defining $f(z_0) = a_0$ we can consider f to be analytic at z_0. (This is a trivial example of extension to an analytic function!)

EXAMPLE 1(A). $f(z) = \dfrac{\sin z}{z} \ (z \neq 0)$

$$= 1 - \frac{z^2}{3!} + \frac{z^4}{5!} - \ldots + \frac{(-1)^n z^{2n}}{(2n+1)!} + \ldots$$

If z_0 is an isolated singularity of f and $\lim_{z \to z_0} f(z)$ is finite, then z_0 must be a removable singularity. This is because

$$f(z) = \sum_{n=0}^{\infty} a_n(z-z_0)^n + \sum_{n=1}^{\infty} b_n(z-z_0)^{-n} \qquad 0 < |z-z_0| < R$$

where $b_n = \dfrac{1}{2\pi i} \displaystyle\int_C (z-z_0)^{n-1} f(z) dz$, C being the circle centre z_0,

28

radius r. But $\lim\limits_{z \to z_0} f(z)$ is finite and so in a neighbourhood of z_0 we have $|f(z)| \leqslant M$ for some M. This gives $|b_n| \leqslant \dfrac{1}{2\pi} r^{n-1} M 2\pi r = M r^n$ and letting $r \to 0$, we see that $b_n = 0$ for $n \geqslant 1$.

EXAMPLE 1(B). $f(z) = \dfrac{z}{e^z - 1}$ has a removable singularity at the origin, because as $z \to 0$, we have

$$\frac{e^z - 1}{z} = 1 + \frac{z}{2!} + \ldots + \frac{z^{n-1}}{n!} + \ldots \to 1$$

and so $f(z) \to 1$.

CASE 2

The principal part is a finite series, $b_m \neq 0$ but $b_n = 0$ for $n > m$. In this case we call z_0 a *pole of order m*. A pole of order $1, 2, 3, \ldots$ is also termed simple, double, triple, $\ldots$ respectively. For a pole of order m we have

$$f(z) = \frac{b_m}{(z-z_0)^m} + \ldots + \frac{b_1}{z-z_0} + \sum_{n=0}^{\infty} a_n (z-z_0)^n \quad 0 < |z-z_0| < R$$

$$= (z-z_0)^{-m} g(z)$$

where

$$g(z) = b_m + b_{m-1}(z-z_0) + \ldots + b_1(z-z_0)^{m-1} + \sum_{n=0}^{\infty} a_n (z-z_0)^{m+n}$$

is analytic for $|z-z_0| < R$ and $g(z_0) = b_m \neq 0$.

This implies that $g(z) \neq 0$ in a small neighbourhood of z_0 and since $\dfrac{1}{f(z)} = \dfrac{(z-z_0)^m}{g(z)}$, we see that $\dfrac{1}{f}$ has a zero of order m at z_0.

Hence as $z \to z_0$ we have $\dfrac{1}{|f(z)|} \to 0$ and so $|f(z)| \to +\infty$.

EXAMPLE 2(A). $f(z) = \dfrac{1}{z^2-1}$ $(z \neq 1)$.

Put $z = 1+h$, then

$$f(z) = \frac{1}{h(2+h)}$$

$$= \frac{1}{2h}\{1 - \tfrac{1}{2}h + \tfrac{1}{4}h^2 - \ldots + (-\tfrac{1}{2})^n h^n + \ldots\} \text{ for } 0 < |h| < 2$$

$$= \frac{1}{2h} - \tfrac{1}{4} + \frac{h}{8} - \ldots - (-\tfrac{1}{2})^{n+2} h^n + \ldots$$

Thus f has a simple pole at $z = 1$.

It is possible to show that a point is a pole of order m without actually calculating the Laurent series. If

$$f(z) = b_m(z-z_0)^{-m} + \ldots + b_1(z-z_0)^{-1}$$
$$+ \sum_{n=0}^{\infty} a_n(z-z_0)^n \qquad 0 < |z-z_0| < R$$

then $(z-z_0)^m f(z) \to b_m \neq 0$ as $z \to z_0$. Conversely, if $(z-z_0)^m f(z)$ tends to a non-zero limit, then, as we have seen, $(z-z_0)^m f(z)$ has a removable singularity at z_0 and so $f(z)$ has a pole of order m. (It may also be seen that $(z-z_0)^n f(z) \to 0$ for $n > m$ and $(z-z_0)^n f(z)$ does not tend to a finite limit for $n < m$.)

EXAMPLE 2(B). $f(z) = \dfrac{2z+4}{(1-z^2)\sin^3 z}$ has a triple pole at the origin because $z^3 f(z) = \dfrac{2z+4}{1-z^2}\left(\dfrac{z}{\sin z}\right)^3 \to 4$ as $z \to 0$.

CASE 3

The principal part is an infinite series, i.e. an infinite number of the b_n are non-zero. Such a singularity is called an *isolated essential singularity*. The behaviour of f near z_0 is very peculiar.

As $z \to z_0$, we cannot have $|f(z)| \to +\infty$ because this would imply that f has a pole at z_0. (This follows because $|f(z)| \to +\infty$ implies $\dfrac{1}{f(z)} \to 0$, so $\dfrac{1}{f}$ has a removable singularity at z_0 and may be considered analytic there. Since $\lim\limits_{z \to z_0} \dfrac{1}{f(z)} = 0$, $\dfrac{1}{f}$ has a zero of order m for some $m \geqslant 1$, and f must have a pole of order m.)

If $f(z)$ does not approach infinity, what happens? In fact the behaviour of f is very wild near z_0 in the sense that in any neighbourhood of z_0 (however small) f takes every complex value with perhaps one exception. This is Picard's Theorem; the proof is omitted.

EXAMPLE 3. $\exp(1/z) = 1 + \dfrac{1}{z} + \dfrac{1}{2!z^2} + \ldots + \dfrac{1}{n!z^n} + \ldots \; |z| > 0.$

In $0 < |z| < \varepsilon$ (no matter how small ε), $\exp(1/z)$ takes on every complex value except $w = 0$. To see this, we require to find z such that $w = \exp(1/z)$, $0 < |z| < \varepsilon$. This is equivalent to solving the equations:

(a) $\quad \dfrac{1}{z} = \text{Log}|w| + i(\arg w + 2\pi k)$ $\quad$ (b) $\quad \dfrac{1}{|z|^2} > \dfrac{1}{\varepsilon^2}.$

For $w \neq 0$ and any integer k we can find z from (a), and by choosing k very large, we can make

$$\frac{1}{|z|^2} = (\text{Log } |w|)^2 + (\arg w + 2\pi k)^2 > \frac{1}{\varepsilon^2}.$$

Note. If $z_1, z_2, \ldots$ is a sequence of distinct isolated singularities of f which tends to a limit z_0, then z_0 cannot be an isolated singularity of f. This is because every annulus $0 < |z - z_0| < \varepsilon$ contains points of the sequence and at these points f is not analytic. In such a case, z_0 is called an *essential singularity* of f.

EXAMPLE 4. $f(z) = \left(\sin\left(\dfrac{1}{z} \right) \right)^{-1}$ has an essential singularity

at the origin because $\dfrac{1}{\pi}, \dfrac{1}{2\pi}, \ldots, \dfrac{1}{n\pi}, \ldots$ is a sequence of

singularities of f which tends to zero.

3. The Point at Infinity

In the last section we saw that if z_0 was a pole of f, then $|f(z)| \to +\infty$ as $z \to z_0$. It is possible to adjoin a single point at infinity (denoted by ∞) to the complex plane so that $f(z) \to \infty$ as $z \to z_0$.

Consider a sphere touching the complex plane at the origin and let N (the 'north pole') be the point on the sphere diametrically opposite the point of contact.

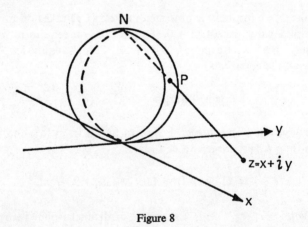

Figure 8

If P is any point on the sphere distinct from N, then the straight line NP meets the plane in a unique point $z = x + iy$ and this sets up a correspondence between all the points of the sphere except N and all the points of the complex plane.

We note that N is omitted in this correspondence and we suppose that it corresponds to the symbol ∞. The complex plane together with ∞ is called the *extended complex plane* and we see that there is a correspondence between the points on the sphere and the points of the extended complex plane.

We remark that 'lines of latitude' on the sphere correspond to circles of the form $|z| = R$ in the plane and the 'polar cap' between a line of latitude and the north pole corresponds to the domain $|z| > R$. As R increases, the corresponding line of latitude approaches N. For this reason we define the domain $|z| > R$ (together with ∞) to be a neighbourhood of ∞ and we write $w \to \infty$ if the real number $|w| \to +\infty$. For example if z_0 is a pole of f, then $\lim\limits_{z \to z_0} f(z) = \infty$.

Suppose that f is analytic for $|z| > R$, then $\dfrac{d}{dz}\left(f\left(\dfrac{1}{z}\right)\right) = -\dfrac{1}{z^2} f'\left(\dfrac{1}{z}\right)$ and so $f\left(\dfrac{1}{z}\right)$ is analytic for $\left|\dfrac{1}{z}\right| > R$, i.e. for $0 < |z| < \dfrac{1}{R}$. Thus $f\left(\dfrac{1}{z}\right)$ has an isolated singularity at the origin. We say that $f(z)$ has a removable singularity at ∞, pole of order m at ∞, or isolated essential singularity at ∞ if $f\left(\dfrac{1}{z}\right)$ has the corresponding singularity at the origin. In particular, if f has a removable singularity at ∞, we may regard f as being analytic at ∞ and define $f(\infty) = \lim\limits_{z \to \infty} f(z) = \lim\limits_{z \to 0} f\left(\dfrac{1}{z}\right)$.

EXAMPLE 1. $f(z) = z^{-3} \exp\left(\dfrac{1}{z}\right)$ has a removable singularity at ∞ since $f\left(\dfrac{1}{z}\right) = z^3 e^z$ $(z \neq 0)$ has a removable singularity at the origin.

EXAMPLE 2. $f(z) = z^3$ has a triple pole at ∞ since $f\left(\dfrac{1}{z}\right) = \dfrac{1}{z^3}$.

EXAMPLE 3. $f(z) = e^z$ has an isolated essential singularity at ∞ since $f\left(\dfrac{1}{z}\right) = \exp\left(\dfrac{1}{z}\right)$.

If $z_1, z_2, \ldots$ is a sequence of isolated singularities of f and $\lim_{n\to\infty} z_n = \infty$, then f cannot have an isolated singularity at ∞ since every domain $|z| > R$ contains points of the sequence where f is not analytic. In this case f is said to have an essential singularity at ∞.

EXAMPLE 4. $f(z) = \tan z$ has an essential singularity at ∞ since $(n+\tfrac{1}{2})\pi$ is a singularity of f for every integer n.

4. Cauchy's Residue Theorem

If z_0 is an isolated singularity of f, then by Laurent's Theorem we have

$$f(z) = \sum_{n=0}^{\infty} a_n(z-z_0)^n + \sum_{n=1}^{\infty} b_n(z-z_0)^{-n} \quad 0 < |z-z_0| < R.$$

Also the coefficient b_n is given by

$$b_n = \frac{1}{2\pi i}\int_C (z-z_0)^{n-1} f(z)dz$$

where C is the circle centre z_0, radius r, $z(t) = z_0 + re^{it}$ $(0 \leqslant t \leqslant 2\pi)$. In particular, the case $n = 1$ holds a special place because

$$b_1 = \frac{1}{2\pi i}\int_C f(z)dz.$$

The coefficient b_1 is called the *residue* of f at z_0.

(Note that the importance of b_1 is to be expected, for term by term integration gives

$$\int_C f(z)dz = \int_C \left\{ \sum_{n=0}^{\infty} a_n(z-z_0)^n + \sum_{n=1}^{\infty} b_n(z-z_0)^{-n} \right\} dz$$

$$= \sum a_n \int_C (z-z_0)^n dz + \sum b_n \int_C (z-z_0)^{-n} dz$$

$$= b_1.2\pi i.$$

The last line follows because $(z-z_0)^n = \dfrac{d}{dz}\left\{ \dfrac{(z-z_0)^{n+1}}{n+1} \right\}$ for $n \neq -1$, giving $\int_C (z-z_0)^n dz = 0$ $(n \neq -1)$, whereas $\int_C (z-z_0)^{-1}dz = 2\pi i$ by direct calculation.)

Suppose γ is a closed Jordan contour (described anti-clockwise) whose track lies in the domain of definition of f and suppose f is analytic everywhere inside γ except at the isolated singularity z_0.

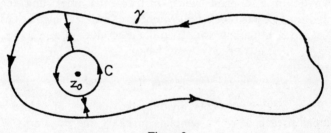

Figure 9

By choosing a small circle around z_0, making cuts from γ to C in the usual fashion, we see that

$$b_1 = \frac{1}{2\pi i} \int_C f(z)dz = \frac{1}{2\pi i} \int_\gamma f(z)dz.$$

Hence if we know the residue b_1 of f at z_0, we can calculate $\int_\gamma f(z)dz$ by the formula

$$\int_\gamma f(z)dz = 2\pi i b_1 \qquad (2)$$

EXAMPLE. $f(z) = \dfrac{1}{z}$ has residue 1 at the origin. Hence if γ is any closed Jordan contour described anti-clockwise round the origin,

$$\int_\gamma \frac{1}{z}\, dz = 2\pi i$$

This generalizes the case where γ is the unit circle which may be calculated directly.

This method of calculating integrals by residues is an extremely useful technique. It generalizes to the case of several singularities inside γ.

CAUCHY'S RESIDUE THEOREM. Let γ be a closed Jordan contour described anti-clockwise. Suppose the function f is analytic in a domain which includes the track and the interior of γ except for a finite number of isolated singularities $z_1, \ldots, z_n$ in the interior. Then if the residues at $z_1, \ldots, z_n$ are $\rho_1, \ldots, \rho_n$ respectively we have

$$\int_\gamma f(z)\, dz = 2\pi i(\rho_1 + \ldots + \rho_n).$$

Proof. Make cross-cuts dividing the interior of γ into n domains, each of which contains precisely one singularity.

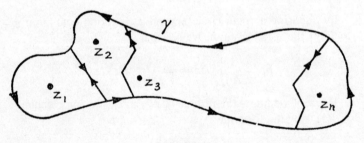

Figure 10

If Γ_r is the boundary contour (described anti-clockwise) of the region containing z_r, by (2) we have

$$\int_{\Gamma_r} f(z)dz = 2\pi i \rho_r.$$

Adding these integrals, the contributions due to the cross-cuts cancel in pairs and we find

$$\int_\gamma f(z)dz = 2\pi i(\rho_1 + \ldots + \rho_n).$$

Note. This is yet another proof which relies on geometric intuition because we have not specified precisely how to make the cuts. Nevertheless, in any particular case that we meet in this text, this would be obvious.

In Chapter Three we will use Cauchy's Residue Theorem to calculate specific integrals and will give several examples there. We now use the theorem to obtain some more general results.

5. Number of Zeros and Poles

Cauchy's Residue Theorem may be used to find the number of zeros and poles of an analytic function inside a closed Jordan contour. For this purpose a zero of order m is counted m times and a pole of order n is counted n times.

THEOREM 5.1. Let γ be a closed Jordan contour described anti-clockwise. Suppose that f is analytic in a domain which includes the track and interior of γ except possibly for a finite number of poles inside γ. If f is non-zero on the track of γ, then

$$\frac{1}{2\pi i}\int_\gamma \frac{f'(z)}{f(z)} dz = N - P$$

where N is the number of zeros and P is the number of poles inside γ.

Proof. First note that the integral is well-defined because f is non-zero on the track of γ and so $\dfrac{f'}{f}$ is analytic there. In fact $\dfrac{f'}{f}$ only has poles where f has a zero or f' (and hence f) has a pole.

If z_0 is a zero of order m, we have $f(z) = (z-z_0)^m g(z)$ where g is analytic and non-zero in a neighbourhood of z_0. Thus

$$f'(z) = m(z-z_0)^{m-1}g(z) + (z-z_0)^m g'(z)$$

$$\text{and} \quad \frac{f'(z)}{f(z)} = \frac{m}{z-z_0} + \frac{g'(z)}{g(z)}.$$

But $\dfrac{g'}{g}$ is analytic in the neighbourhood of z_0 and so $\dfrac{f'}{f}$ has a simple pole of residue m at z_0.

Similarly if f has a pole of order n at z_1, then $f(z) = (z-z_1)^{-n}h(z)$ where h is analytic and non-zero in a neighbourhood of z_1. Thus

$$f'(z) = -n(z-z_1)^{-n-1}h(z) + (z-z_1)^{-n}h'(z)$$

$$\text{and} \quad \frac{f'(z)}{f(z)} = \frac{-n}{z-z_1} + \frac{h'(z)}{h(z)}.$$

Again $\dfrac{h'}{h}$ is analytic in a neighbourhood of z_1 and $\dfrac{f'}{f}$ has a simple pole of residue $-n$ at z_1. By adding all the residues together, we obtain the required result.

As a corollary of this theorem, we see that if f is actually analytic inside γ, then the number of zeros of f inside γ is

$$\frac{1}{2\pi i} \int_\gamma \frac{f'(z)}{f(z)} \, dz.$$

ROUCHÉ'S THEOREM. Suppose that f and g are both

analytic in a domain containing the track and interior of a closed Jordan contour γ (described anti-clockwise). If $|g(z)| < |f(z)|$ on the track of γ then f and $f+g$ have the same number of zeros inside γ.

Proof. Suppose that f has m zeros and $f+g$ has n zeros inside γ. Then if $F(z) = \dfrac{f(z)+g(z)}{f(z)}$, we see that F has n zeros and m poles inside γ. Also $|f(z)| > |g(z)| \geqslant 0$ on the track of γ showing that F is analytic there.

We will show

$$n - m = \frac{1}{2\pi i} \int_\gamma \frac{F'(z)}{F(z)} \, dz = 0.$$

This is done by transforming the integral.

Write $w = F(z)$, then as z describes the contour γ in the z-plane, w describes a contour Γ in the w-plane. Explicitly, if γ is given by $z(t) = x(t) + iy(t)$ $(\alpha \leqslant t \leqslant \beta)$, then Γ is given by $w(t) = F(z(t))$ $(\alpha \leqslant t \leqslant \beta)$.
Hence

$$\int_\gamma \frac{F'(z)}{F(z)} \, dz = \int_\alpha^\beta \frac{F'(z(t))}{F(z(t))} \, z'(t) dt$$

$$= \int_\alpha^\beta \frac{w'(t)}{w(t)} \, dt$$

$$= \int_\Gamma \frac{1}{w} \, dw.$$

But for w on the track of Γ, the real part satisfies

$$\Re w = \Re(F(z)) = \Re\left(\frac{f(z)+g(z)}{f(z)} \right)$$

$$= 1 + \Re\left(\frac{g(z)}{f(z)} \right) \geqslant 1 - \left| \frac{g(z)}{f(z)} \right| > 0,$$

by hypothesis.

39

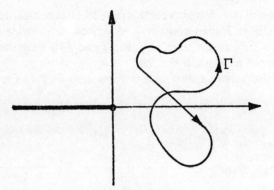

Figure 11

This means that the track of Γ lies in the half-plane $\Re w > 0$ and so must lie in the cut w-plane (cut along the negative real axis).

In the cut-plane we have $\dfrac{1}{w} = \dfrac{d}{dw}(\text{Log } w)$, and by the Fundamental Theorem of Contour Integration round a closed contour, $\displaystyle\int_{\Gamma} \dfrac{1}{w}\, dw = 0$. This completes the proof.

As a consequence of Rouché's Theorem, we can deduce the *Fundamental Theorem of Algebra*. This states that a polynomial equation

$$z^n + a_1 z^{n-1} + \ldots + a_n = 0$$

has n roots (counted according to multiplicity).

Take $f(z) = z^n$, $g(z) = a_1 z^{n-1} + \ldots + a_n$. Let C be the circle centre the origin, radius $R \geqslant 1$. On C we have $|f(z)| = R^n$ and

$$|g(z)| \leqslant |a_1| R^{n-1} + \ldots + |a_n| \leqslant (|a_1| + \ldots + |a_n|) R^{n-1}.$$

Hence choosing $R > |a_1| + \ldots + |a_n|$, we have $|g(z)| < |f(z)|$ on C.

But f has precisely one zero of order n (at the origin) inside

40

C and so $f+g$ has n zeros inside C. Thus the polynomial equation has n solutions.

Notice that we have only shown that the polynomial has n zeros; we have also given their approximate location, inside the circle C. In particular cases we can use Rouché's Theorem to give further information of this kind.

EXAMPLE. $z^9 - 6z^2 + 10 = 0$ has all nine zeros between the circles $|z| = 1$ and $|z| = 2$.

Consider the circle $|z| = 1$, $g(z) = z^9 - 6z^2$, $f(z) = 10$. If $|z| = 1$, then $|g(z)| = |z^9 - 6z^2| \leqslant |z|^9 + 6|z|^2 = 7 < |f(z)|$. Since $f(z)$ has no zeros inside $|z| = 1$, $f(z) + g(z) = z^9 - 6z^2 + 10$ also has no zeros there.

Similarly on $|z| = 2$, $f(z) = z^9$, $g(z) = 10 - 6z^2$, we have

$$|g(z)| \leqslant 10 + 6|z|^2 = 10 + 24 < 2^9 = |f(z)|$$

and since $f(z)$ has a zero of order 9 at the origin, $f(z) + g(z) = z^9 - 6z^2 + 10$ has 9 zeros inside $|z| = 2$.

EXERCISES ON CHAPTER TWO

For each of the isolated singularities in exercises 1–6, calculate the Laurent expansion and state what type of singularity is involved:

1. $z^{-5}e^{z^3}$ at $z = 0$ 2. $(z^2 - a^2)^{-1}$ at $z = a\,(a > 0)$

3. $z^{-1}\cos(z^{-1})$ at $z = 0$ 4. $(1 - z)e^{1/z}$

5. $z^{-5}(2\cos z + z^2 - 2)$ at $z = 0$ 6. $\{(z-1)(z-2)\}^{-1}$ at $z = 1$.

Classify the singularities of the functions given by the formulae in exercises 7–11 (a) at the origin, (b) at ∞.

7. $\dfrac{e^z}{z \sin^2 z}$ 8. $\dfrac{z}{1 - \cos z}$ 9. $z^3 \sin(z^{-1})$ 10. $\tan(z^{-1})$

11. $z^{-3}e^z$.

12. Use Rouché's Theorem to show that if $|\alpha| > e$, then $\alpha z^n = e^z$ has n solutions inside $|z| = 1$.

CHAPTER THREE

The Calculus of Residues

---

1. Residues

In this chapter we intend to use Cauchy's Residue Theorem to calculate specific integrals. In order to do this we must be able to calculate the residue at an isolated singularity. The most direct method is to calculate part of the Laurent series of f at the singularity z_0 to find the coefficient of $\dfrac{1}{z-z_0}$. In simple cases this calculation may be avoided.

METHOD 1.

For a simple pole, the residue of f at z_0 is $\lim_{z \to z_0} (z-z_0)f(z)$. This is because $f(z) = \dfrac{b_1}{z-z_0} + \sum_{n=0}^{\infty} a_n(z-z_0)^n$ and so b_1 is the given limiting value.

EXAMPLE 1. If $f(z) = \dfrac{z}{1-\cos z}$, then the residue of f at zero is

$$\lim_{z \to 0} z \frac{z}{1-\cos z} = \lim_{z \to 0} \frac{4(\tfrac{1}{2}z)^2}{2\sin^2(\tfrac{1}{2}z)} = 2.$$

Sometimes we have $f(z) = \dfrac{p(z)}{q(z)}$ and f has a pole at z_0 because $q(z)$ is zero there.

METHOD 2.

If $f(z) = \dfrac{p(z)}{q(z)}$ where $p(z_0) \neq 0$ and z_0 is a simple zero of q,

then the residue of f at z_0 is $\dfrac{p(z_0)}{q'(z_0)}$.

This is because $q(z_0) = 0$ and $q'(z_0) \neq 0$, hence

$$\lim_{z \to z_0} (z - z_0) f(z) = \lim_{z \to z_0} p(z) \bigg/ \left\{ \frac{q(z) - q(z_0)}{z - z_0} \right\} = \frac{p(z_0)}{q'(z_0)} .$$

EXAMPLE 2. If $f(z) = \dfrac{1}{1 - z^4}$ then the residue of f at $z_0 = 1$

is $\dfrac{1}{-4z_0^3} = -\dfrac{1}{4}$.

Methods 1, 2 may be generalized for poles of higher order, but the calculations sometimes become complicated and then the best method is direct calculation from the Laurent series. However, generalizing Method 1 for a pole of order m, we have:

METHOD 3.

If z_0 is a pole of order m of the function f, then the residue of f at z_0 is

$$\lim_{z \to z_0} \frac{1}{(m-1)!} \frac{d^{m-1}}{dz^{m-1}} \{(z - z_0)^m f(z)\}.$$

This is because

$$f(z) = b_m(z - z_0)^{-m} + \ldots + b_1(z - z_0)^{-1} + \sum_{n=0}^{\infty} a_n(z - z_0)^n,$$

and so

$$(z - z_0)^m f(z) = b_m + \ldots + b_1(z - z_0)^{m-1} + \sum_{n=0}^{\infty} a_n(z - z_0)^{m+n}.$$

This gives

$$\frac{d^{m-1}}{dz^{m-1}}\{(z-z_0)^m f(z)\} = (m-1)!\,b_1 + m!\,a_0(z-z_0) + \ldots$$

and the result follows.

EXAMPLE 3. If $f(z) = \left(\dfrac{z+1}{z-1}\right)^2$ then the residue at the double pole $z_0 = 1$ is

$$\lim_{z \to 1} \frac{1}{1!} \frac{d}{dz}\left\{(z-1)^2 \left(\frac{z+1}{z-1}\right)^2\right\} = \lim_{z \to 1}(2z+2) = 4.$$

In cases where methods 1-3 are not applicable, or the calculations become difficult, we must determine the relevant part of the Laurent series. (We only require the coefficient of $(z-z_0)^{-1}$, so the reader who calculates the whole series is wasting a great deal of energy!)

The calculation can often be performed by manipulating Taylor series. We recall that we can add or multiply power series $\sum\limits_{n=0}^{\infty} a_n(z-z_0)^n$, $\sum\limits_{n=0}^{\infty} b_n(z-z_0)^n$ term by term in any disc $|z-z_0| < R$ where both series converge. In particular the product is $\sum\limits_{n=0}^{\infty} c_n(z-z_0)^n$ where $c_n = a_0 b_n + a_1 b_{n-1} + \ldots + a_n b_0$.

To calculate $1/f(z)$ where $f(z) = \sum\limits_{n=0}^{\infty} a_n(z-z_0)^n$ for $|z-z_0| < R$ and $a_0 \neq 0$, we remark first that $1/f(z)$ certainly has a unique power series expansion $\sum\limits_{n=0}^{\infty} b_n(z-z_0)^n$ in a small disc centre z_0. (Because $f(z_0) = a_0 \neq 0$ and by continuity $f(z) \neq 0$ in $|z-z_0| < \varepsilon$ for some $\varepsilon > 0$. Hence the inverse $1/f(z)$ is analytic (with derivative $-f'(z)/(f(z))^2$) in $|z-z_0| < \varepsilon$ and so has a unique Taylor series.†) Since $\sum\limits_{n=0}^{\infty} a_n(z-z_0)^n \sum\limits_{n=0}^{\infty} b_n(z-z_0)^n = 1$, multi-

† *Functions of a Complex Variable I*, p. 55-6.

plying out and comparing coefficients we have $a_0 b_0 = 1$, $a_0 b_1 + a_1 b_0 = 0, \ldots, a_0 b_n + \ldots + a_n b_0 = 0, \ldots$. But $a_0 \neq 0$ and so we can use these equations successively to find b_0, $b_1, \ldots$. For example if $f(z) = \dfrac{\sin z}{z} = 1 - \dfrac{z^2}{3!} + \ldots$, then $1/f(z) = b_0 + b_1 z + b_2 z^2 + \ldots$ where

$$(1 - \tfrac{1}{6} z^2 + \ldots)(b_0 + b_1 z + b_2 z^2 + \ldots) = 1.$$

Hence $b_0 = 1$, $b_1 = 0$, $b_2 = \tfrac{1}{6}, \ldots$, implying

$$z/\sin z = 1 + \tfrac{1}{6} z^2 + \text{higher order terms.}$$

We now calculate a residue which will later prove useful.

EXAMPLE 4. The residue of $z^{-2} \cot \pi z$ at the origin.
Replacing z by πz in the series for $z/\sin z$, we find

$$\pi z / \sin \pi z = 1 + \tfrac{1}{6} \pi^2 z^2 + \ldots$$

Hence $z^{-2} \cot \pi z = \dfrac{1}{\pi z^3} \cos \pi z \dfrac{\pi z}{\sin \pi z}$

$$= \dfrac{1}{\pi z^3}(1 - \tfrac{1}{2}\pi^2 z^2 + \ldots)(1 + \tfrac{1}{6}\pi^2 z^2 + \ldots).$$

The coefficient of $1/z$ is $\pi(\tfrac{1}{6} - \tfrac{1}{2}) = -\tfrac{1}{3}\pi$, i.e. the residue is $-\tfrac{1}{3}\pi$.

2. Integrals of the Form $\int_0^{2\pi} f(\cos t, \sin t)\, dt$

If C is the unit circle $z(t) = \cos t + i \sin t$ $(0 \leqslant t \leqslant 2\pi)$ we may transform $\int_0^{2\pi} f(\cos t, \sin t)\, dt$ into a contour integral of the form $\int_C g(z)\, dz$ and use Cauchy's Residue Theorem to calculate the latter. This is always possible if the function g is analytic in a domain including C and its interior except possibly for a finite number of isolated singularities inside C.

Specifically, if $z = e^{it}$ then $\cos t = \tfrac{1}{2}\left(z + \dfrac{1}{z}\right)$,

$$\sin t = \frac{1}{2i}\left(z - \frac{1}{z}\right) \text{ and } z'(t) = ie^{it} = iz.$$

Let $g(z) = f\left(\frac{1}{2}\left(z + \frac{1}{z}\right), \frac{1}{2i}\left(z - \frac{1}{z}\right)\right)\frac{1}{iz}$.

then†

$$\int_C g(z)dz = \int_0^{2\pi} g(z(t))z'(t)dt = \int_0^{2\pi} f(\cos t, \sin t)dt.$$

Thus $\int_0^{2\pi} f(\cos t, \sin t)dt = 2\pi i$(sum of residues of g at isolated singularities inside C).

EXAMPLE. $I = \int_0^{2\pi} \frac{dt}{a + b \cos t} \qquad (a > b > 0).$

We find $I = \int_C \frac{1}{a + \frac{1}{2}b(z + 1/z)} \cdot \frac{1}{iz} dz$

$$= \frac{2}{i} \int_C \frac{dz}{bz^2 + 2az + b}$$

$$= \frac{2}{i} \int_C \frac{dz}{q(z)}.$$

Since $\frac{1}{q(z)}$ only has poles where $q(z) = bz^2 + 2az + b = 0$, there are simple poles at $\frac{-a \pm \sqrt{(a^2 - b^2)}}{b}$. Let $\alpha = \frac{-a + \sqrt{(a^2 - b^2)}}{b}$, $\beta = \frac{-a - \sqrt{(a^2 - b^2)}}{b}$, then $\alpha\beta = \frac{b}{b} = 1$ and so $|\alpha||\beta| = 1$.

Since $|\alpha| < |\beta|$, we must have $|\alpha| < 1$, $|\beta| > 1$, and the only pole

---

† The reader may also remember this formula by substituting for $\cos t$, $\sin t$ and $dt = \frac{dz}{iz}$ in $\int_0^{2\pi} f(\cos t, \sin t) dt = \int_C f\left(\frac{1}{2}\left(z + \frac{1}{z}\right), \frac{1}{2i}\left(z + \frac{1}{z}\right)\right)\frac{dz}{iz}$ $= \int_C g(z)dz$, but strictly speaking we have not justified the use of the differential dz as a separate entity.

of $\dfrac{1}{q(z)}$ inside C is a simple pole at α with residue $\dfrac{1}{q'(z)} =$

$$\frac{1}{2b\alpha + 2a} = \frac{1}{2\sqrt{(a^2 - b^2)}}.$$

Hence $I = \dfrac{2}{i} \displaystyle\int_C \dfrac{dz}{q(z)} = 2\pi i \cdot \dfrac{2}{i} \cdot \dfrac{1}{2\sqrt{(a^2 - b^2)}} = \dfrac{2\pi}{\sqrt{(a^2 - b^2)}}.$

3. Integrals of the Form $\int_{-\infty}^{\infty} f(x)\,dx$

Under suitable conditions we have

$\int_{-\infty}^{\infty} f(x)\,dx = 2\pi i$ (sum of residues of f at isolated singularities in the upper half-plane).

To obtain this result, we integrate round the contour composed of the semicircle S_R given by $z(t) = Re^{it}$ $(0 \leqslant t \leqslant 2\pi)$ and its diameter from $-R$ to R.

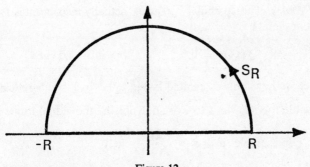

Figure 12

The calculation is possible if:

(i) f is analytic in a domain which includes the upper half-plane ($\mathscr{I} z \geqslant 0$) except for a finite number of isolated singularities which do not lie on the real axis.

(ii) for large R, $|f(z)| \leqslant \dfrac{M}{R^2}$ when z lies on the semicircle S_R.

To see this we choose R so large that (ii) is satisfied and also all the singularities lie inside the closed contour of figure 12.

Then we have

$\int_{-R}^{R} f(x)dx + \int_{S_R} f(z)dz = 2\pi i$ (sum of residues in the upper half-plane).

Now let $R \to \infty$. Since

$$\left| \int_{S_R} f(z)dz \right| \leqslant \frac{M}{R^2} \cdot \pi R = \frac{\pi M}{R}$$

we have $\lim\limits_{R \to \infty} \int_{S_R} f(z)dz = 0$.

Thus

$$\lim_{R \to \infty} \int_{-R}^{R} f(x)dx = 2\pi i \text{ (sum of residues in upper half-plane)}.$$

Remark. The symbol $\int_{-\infty}^{\infty} f(x)dx$ actually incorporates two distinct limits,

$$\int_{-\infty}^{\infty} f(x)dx = \lim_{Y \to \infty} \int_{-Y}^{0} f(x)dx + \lim_{X \to \infty} \int_{0}^{X} f(x)dx. \qquad (1)$$

Since we have only calculated $\lim\limits_{R \to \infty} \int_{-R}^{R} f(x)dx$, it is theoretically possible for this limit to exist but not the individual limits in

(1). For example, if $\phi(x) = \dfrac{2x}{x^2+1}$, then

$$\int_{-Y}^{X} \phi(x)dx = \log\left(\frac{X^2+1}{Y^2+1} \right).$$

Thus we find that $\int_{-R}^{R} \phi(x) = 0$ and $\lim\limits_{R \to \infty} \int_{-R}^{R} \phi(x)dx = 0$, but $\lim\limits_{Y \to \infty} \int_{-Y}^{0} \phi(x)dx = -\infty$, and $\lim\limits_{X \to \infty} \int_{0}^{X} \phi(x)dx = +\infty$. In such a case, $\lim_{R \to \infty} \int_{-R}^{R} \phi(x) dx$ is called the *Cauchy principal value* and is denoted by $P\int_{-\infty}^{\infty} \phi(x) dx$. Luckily $\phi(x) = \dfrac{2x}{x^2 + 1}$ does not satisfy condition (ii) and subject to this condition, there is no problem with the limits. This is because there is a comparison test for infinite integrals analogous to the real case.†

If $p(x)$ is a continuous, positive real-valued function such that $|f(x)| \leqslant p(x)$ for $x \geqslant K$ and $\lim_{X \to \infty} \int_{K}^{X} p(x) dx$ exists, then $\lim_{X \to \infty} \int_{K}^{X} f(x) dx$ exists. (To prove this, note that $|\Re f(x)| \leqslant |f(x)| \leqslant |p(x)|$ and so $\lim_{X \to \infty} \int_{K}^{X} \Re f(x) dx$ exists by the comparison test in the real case; similarly for the imaginary part.) Using condition (ii) and comparing $|f(x)|$ with $p(x) = \dfrac{M}{x^2}$, we see

$$\int_{K}^{X} \frac{M}{x^2} dx = \frac{M}{K} - \frac{M}{X} \text{ tends to } \frac{M}{K} \text{ as } X \to \infty.$$

Hence $\lim_{X \to \infty} \int_{K}^{X} f(x) dx$ exists and similarly for $\lim_{Y \to \infty} \int_{-Y}^{K} f(x) dx$. Thus $\int_{-\infty}^{\infty} f(x) dx$ exists.

A suitable function for this type of calculation is any rational function $\dfrac{N(z)}{D(z)}$ where N, D are polynomials such that

  (i) $D(x) \neq 0$ when x is real,
  (ii) degree $D \geqslant 2 + $ degree N.

EXAMPLE 1. $\displaystyle\int_{-\infty}^{\infty} \frac{dx}{(x^2 + a^2)(x^2 + b^2)} = \frac{\pi}{ab(a+b)}$ when $a > 0$, $b > 0$, $a \neq b$.

The only singularities of the integrand in the upper half-plane are simple poles at ia, ib. The residue at ia is

$$\lim_{z \to ia} \frac{z - ia}{(z^2 + a^2)(z^2 + b^2)} = \frac{1}{2ia(b^2 - a^2)}$$

† W. Ledermann, *Integral Calculus*, pp. 21, 22.

THE CALCULUS OF RESIDUES

and at ib it is $\dfrac{1}{2ib(a^2-b^2)}$.

Thus
$$\int_{-\infty}^{\infty} \frac{dx}{(x^2+a^2)(x^2+b^2)} = 2\pi i\left(\frac{1}{2ia(b^2-a^2)}+\frac{1}{2ib(a^2-b^2)}\right)$$
$$= \frac{\pi(b-a)}{ab(b^2-a^2)}$$
$$= \frac{\pi}{ab(a+b)} .$$

As a further refinement, note that we did not require $f(z)$ to be real on the real axis. The function e^{imz} $(m>0)$ is everywhere analytic and satisfies

$$|e^{imz}| = |e^{imx-my}| = |e^{-my}| \leqslant 1 \text{ for } y \geqslant 0 \text{ (since } m>0).$$

Hence if $f(z)$ satisfies conditions (i), (ii), then so does $e^{imz}f(z)$.

EXAMPLE 2. Consider $f(z) = \dfrac{1}{(z^2+a^2)(z^2+b^2)}$ where $a>0$, $b>0$, $a \neq b$.

The residue of $e^{imz}f(z)$ at ia is

$$\lim_{z \to ia} \frac{(z-ia)e^{imz}}{(z^2+a^2)(z^2+b^2)} = \frac{e^{-ma}}{2ia(b^2-a^2)}$$

and at ib it is $\dfrac{e^{-mb}}{2ib(a^2-b^2)}$. Thus we have

$$\int_{-\infty}^{\infty} \frac{e^{imx}}{(x^2+a^2)(x^2+b^2)} dx = 2\pi i\left(\frac{e^{-ma}}{2ia(b^2-a^2)}+\frac{e^{-mb}}{2ib(a^2-b^2)}\right)$$

Equating real and imaginary parts, this gives

$$\int_{-\infty}^{\infty} \frac{\cos mx}{(x^2+a^2)(x^2+b^2)} dx = \frac{\pi}{b^2-a^2}\left(\frac{e^{-ma}}{a}-\frac{e^{-mb}}{b}\right)$$

$$\int_{-\infty}^{\infty} \frac{\sin mx}{(x^2+a^2)(x^2+b^2)}\, dx = 0.$$

Notice that if $g(x)$ is an odd function $(g(-x) = -g(x))$, as in the second case, then we must have $\int_{-\infty}^{\infty} g(x)dx = 0$. Also if $g(x)$ is even $(g(-x) = g(x))$, then $\int_{-\infty}^{\infty} g(x)dx = 2\int_{0}^{\infty} g(x)dx$. Thus from example 1,

$$\int_{0}^{\infty} \frac{dx}{(x^2+a^2)(x^2+b^2)} = \frac{\pi}{2ab(a+b)}$$

and from example 2,

$$\int_{0}^{\infty} \frac{\cos mx}{(x^2+a^2)(x^2+b^2)}\, dx = \frac{\pi}{2(b^2-a^2)}\left(\frac{e^{-ma}}{a} - \frac{e^{-mb}}{b}\right).$$

4. Integrals of the Form $\int_{-\infty}^{\infty} e^{imx} f(x)dx$

Integrals of this form are substantially covered by the conditions of the last section. However we can make a slight improvement in condition (ii) below.

For $m > 0$, we have $\int_{-\infty}^{\infty} e^{imx} f(x)dx = 2\pi i$ (sum of residues of $e^{imz}f(z)$ at isolated singularities in the upper half-plane) provided that

(i) f is analytic in a domain containing the upper half-plane except for a finite number of isolated singularities, none of which lie on the real axis.

(ii) for large R, $|f(z)| \leqslant \dfrac{M}{R}$ when $|z| = R$, $\mathscr{I}z \geqslant 0$.

We may use a semicircular contour† as in the last section and prove that $\int_{S_R} e^{imz}f(z)dz \to 0$ as $R \to \infty$. However this

† This method is used in E. G. Phillips, *Functions of a Complex Variable*, Oliver & Boyd, p. 123.

51

method has a basic drawback: it only calculates

$$\lim_{R \to \infty} \int_{-R}^{R} e^{imx} f(x) dx$$

and we still have to show that

$$\int_{-\infty}^{\infty} e^{imx} f(x) dx$$

exists. This would require a delicate argument. The comparison test is of no use because we only have $|e^{imx} f(x)| \leqslant \dfrac{M}{X}$ and $\int_{K}^{\infty} \dfrac{M}{x} dx$ diverges.

A much better method is to replace the semicircular contour by the rectangular contour in figure 13:

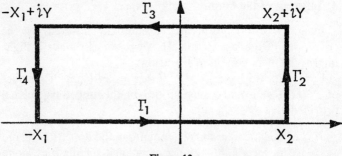

Figure 13

Initially we choose the rectangle large enough to contain all the singularities and such that $|f(z)| \leqslant \dfrac{M}{|z|}$ on Γ_2, Γ_3, Γ_4. If we show that $\int_{\Gamma_2}$, $\int_{\Gamma_3}$, $\int_{\Gamma_4}$ tend to zero, then

$$\lim_{X_1, X_2 \to \infty} \int_{-X_1}^{X_2} e^{imx} f(x) dx = 2\pi i \text{ (sum of residues of } e^{imz} f(z)$$
$$\text{in upper half plane).}$$

In particular, allowing X_1 and X_2 tend to ∞ independently, we know that $\int_{-\infty}^{\infty} e^{imx} f(x)dx$ exists.

$$\left| \int_{\Gamma_2} e^{imz} f(z)dz \right| = \left| \int_0^Y e^{imX_2 - mt} f(X_2 + it)idt \right| \leqslant \int_0^Y e^{-mt} \frac{M}{X_2} dt \leqslant \frac{M}{X_2}$$

and similarly $\left| \int_{\Gamma_4} e^{imz} f(z)dz \right| \leqslant \dfrac{M}{X_1}$.

$$\left| \int_{\Gamma_3} e^{imz} f(z)dz \right| = \left| -\int_{-X_1}^{X_2} e^{imt - mY} f(t + iY)dt \right| \leqslant \int_{-X_1}^{X_2} e^{-mY} \frac{M}{Y} dt$$

$$\leqslant \frac{e^{-mY}}{Y} M(X_1 + X_2).$$

For fixed X_1, X_2, let $Y \to \infty$, then $\dfrac{e^{-mY}}{Y} \to 0$ and so $\int_{\Gamma_3} \to 0$.
Now let X_1, $X_2 \to \infty$ then $\int_{\Gamma_2}$, $\int_{\Gamma_4} \to 0$, giving the required result.

EXAMPLE. $I = \displaystyle\int_{-\infty}^{\infty} \frac{xe^{imx}}{x^2 + a^2} dx \qquad (a > 0, \ m > 0)$

The only singularity of the integrand in the upper half-plane is a simple pole at ia with residue

$$\lim_{z \to ia} \frac{(z - ia)ze^{imz}}{z^2 + a^2} = \frac{iae^{-ma}}{2ia} = \frac{1}{2} e^{-ma}.$$

Hence $I = 2\pi i \cdot \frac{1}{2} e^{-ma} = \pi i e^{-ma}$.
Taking real and imaginary parts

$$\int_{-\infty}^{\infty} \frac{x \cos mx}{x^2 + a^2} dx = 0.$$

$$\int_{-\infty}^{\infty} \frac{x \sin mx}{x^2 + a^2} dx = \pi e^{-ma}.$$

53

Since the second integrand is even, we have

$$\int_0^\infty \frac{x \sin mx}{x^2 + a^2}\, dx = \tfrac{1}{2}\pi e^{-ma} \text{ (where } a > 0, m > 0 \text{ in each integral).}$$

5. Poles on the Real Axis

The methods of sections 3, 4 may be extended to the case where f has poles on the real axis. To accommodate these poles, we draw a small semicircle bypassing each of them and let the radius of each semicircle tend to zero. For example, if f has a pole at the origin, we integrate around one of the contours in figure 14.

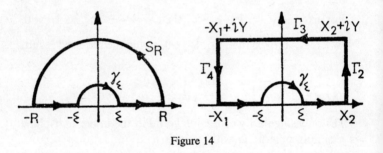

Figure 14

Letting $\varepsilon \to 0$ leads to the same problem as letting $R \to \infty$ in the previous sections. If f has a pole at x_0 where $a \leqslant x_0 \leqslant b$, define the Cauchy principal value of $\int_a^b f(x)dx$ to be

$$P\int_a^b f(x)dx = \lim_{\varepsilon \to 0}\left\{\int_a^{x_0-\varepsilon} f(x)dx + \int_{x_0+\varepsilon}^b f(x)dx\right\}.$$

It may happen that $P\int_a^b f(x)dx$ exists but $\int_a^b f(x)dx$ does not. For example $P\displaystyle\int_{-1}^1 \frac{1}{x}\, dx = 0$.

The above method of contour integration gives the Cauchy

principal value; we must then discuss the convergence of the integral.

EXAMPLE. $\displaystyle\int_{-\infty}^{\infty} \frac{e^{imx}}{x} \, dx \; (m > 0).$

Using the second contour of figure 14, we find $\int_{\Gamma_2}$, $\int_{\Gamma_3}$, $\int_{\Gamma_4} \to 0$ and the integral along the real axis converges at infinity as in section 4. There are no poles of $\dfrac{e^{imz}}{z}$ inside the contour and so

$$\int_{-\infty}^{\varepsilon} \frac{e^{imx}}{x} \, dx + \int_{\varepsilon}^{\infty} \frac{e^{imx}}{x} \, dz + \int_{\gamma_\varepsilon} \frac{e^{mz}}{z} \, dz = 0 \qquad (1)$$

where γ_ε is the opposite contour to $z(t) = e^{it}$ $(0 \leqslant t \leqslant \pi)$ (i.e. γ_ε is the semicircle radius ε, described in the clockwise sense).

But $\dfrac{e^{imz}}{z} = \dfrac{1}{z} + \displaystyle\sum_{n=1}^{\infty} \frac{i^n m^n z^{n-1}}{n!} = \dfrac{1}{z} + g(z)$ where g is analytic and hence $g(z)$ is bounded by M, say, in a neighbourhood of zero. This gives $\left| \int_{\gamma_\varepsilon} g(z) dz \right| \leqslant M\pi\varepsilon$ and so

$$\lim_{\varepsilon \to 0} \int_{\gamma_\varepsilon} \frac{e^{imz}}{z} \, dz = \lim_{\varepsilon \to 0} \int_{\gamma_\varepsilon} \frac{1}{z} \, dz + \lim_{\varepsilon \to 0} \int_{\gamma_\varepsilon} g(z) dz$$

$$= \lim_{\varepsilon \to 0} \left\{ -\int_0^\pi \frac{1}{\varepsilon e^{it}} i\varepsilon e^{it} dt \right\} + 0$$

$$= -i\pi.$$

Thus from equation (1)

$$P\int_{-\infty}^{\infty} \frac{e^{imx}}{x} \, dx = i\pi.$$

Equating real and imaginary parts,

$$P\int_{-\infty}^{\infty} \frac{\cos mx}{x} \, dx = 0, \; P\int_{-\infty}^{\infty} \frac{\sin mx}{x} \, dx = \pi.$$

The first integral only exists as a Cauchy principal value because near zero $\dfrac{\cos mx}{x}$ behaves like $\dfrac{1}{x}$.

$$\text{But } P\int_{-\infty}^{\infty} \frac{\sin mx}{x}\,dx = \lim_{\varepsilon \to 0}\left\{\int_{-\infty}^{-\varepsilon} \frac{\sin mx}{x}\,dx + \int_{\varepsilon}^{\infty} \frac{\sin mx}{x}\,dx\right\}$$

$$= 2\lim_{\varepsilon \to 0}\int_{\varepsilon}^{\infty} \frac{\sin mx}{x}\,dx.$$

Thus $\displaystyle\int_{0}^{\infty} \frac{\sin mx}{x}\,dx$ exists and equals† $\dfrac{\pi}{2}$. This also implies that $\displaystyle\int_{-\infty}^{\infty} \frac{\sin mx}{x}\,dx$ exists and equals π.

Note that the value of $\displaystyle\int_{0}^{\infty} \frac{\sin mx}{x}\,dx$ is independent of the value of m, provided that m is positive. (Compare this result with the example of the previous section as $a \to 0$.)

Clearly we have

$$\int_{0}^{\infty} \frac{\sin mx}{x}\,dx = \begin{cases} \dfrac{\pi}{2} & (m>0) \\[2mm] 0 & (m=0) \\[2mm] -\dfrac{\pi}{2} & (m<0) \end{cases}$$

This result is sometimes called *Dirichlet's discontinuous factor*.

6. Integrals using Periodic Functions

We can use the fact that e^z is periodic, satisfying $e^z = e^{z+2\pi i}$, to calculate certain integrals. We illustrate this with a particular case.

---

† By comparing this proof with one avoiding contour integration, the reader may see the power and elegance of this method. See W. Ledermann, *Integral Calculus*, p. 22, Example 6; p. 37, Example 5.

EXAMPLE. $\int_{-\infty}^{\infty} \dfrac{e^{ax}}{e^x+1}\, dx = \dfrac{\pi}{\sin \pi a} \qquad (0 < a < 1).$

Let $f(z) = \dfrac{e^{az}}{e^z+1}$ and integrate f around the contour in figure 15:

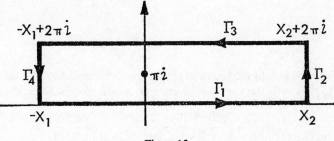

Figure 15

Note that

$$\int_{\Gamma_1} f(z)dz = \int_{-X_1}^{X_2} \frac{e^{ax}}{e^x+1}\, dx \qquad (1)$$

and since Γ_3 is the opposite contour to $z(t) = t+2\pi i$ $(-X_1 \leqslant t \leqslant X_2)$, we have

$$\int_{\Gamma_3} f(z)dz = -\int_{-X_1}^{X_2} \frac{e^{a(t+2\pi i)}}{e^{t+2\pi i}+1}\, dt = -e^{2\pi a i}\int_{-X_1}^{X_2} \frac{e^{ax}}{e^x+1}\, dx \quad (2)$$

Since f has only one singularity inside the rectangular contour, a simple pole at πi with residue $\dfrac{e^{a\pi i}}{e^{\pi i}} = -e^{i\pi a}$, we have

$$\int_{\Gamma_1} f(z)dz + \int_{\Gamma_2} f(z)dz + \int_{\Gamma_3} f(z)dz + \int_{\Gamma_4} f(z)dz = -2\pi i e^{i\pi a}.$$

Let $X_1 \to \infty$, $X_2 \to \infty$, then assuming $\int_{\Gamma_2} \to 0$, $\int_{\Gamma_4} \to 0$, we have from (1), (2)

$$(1-e^{2\pi ai})\int_{-\infty}^{\infty} \frac{e^{ax}}{e^x+1}\,dx = -2\pi i e^{i\pi a}$$

i.e.

$$\int_{-\infty}^{\infty} \frac{e^{ax}}{e^x+1}\,dx = \frac{-2\pi i e^{i\pi a}}{1-e^{2i\pi a}}$$

$$= \frac{2\pi i}{e^{i\pi a}-e^{-i\pi a}}$$

$$= \frac{\pi}{\sin \pi a}.$$

Thus to obtain the required result, it only remains to show that $\int_{\Gamma_2}, \int_{\Gamma_4} \to 0$. But on Γ_2 we have $z = X_2 + it$ $(0 \leqslant t \leqslant 2\pi)$ and so

$$|f(z)| = \frac{|e^{a(X_2+it)}|}{|e^{X_2+it}+1|} = \frac{e^{aX_2}}{|e^{X_2+it}+1|} \leqslant \frac{e^{aX_2}}{e^{X_2}-1}$$

(since $|e^{X_2+it}+1| \geqslant |e^{X_2+it}|-1 = e^{X_2}-1$).

This gives $\left| \int_{\Gamma_2} f(z)dz \right| \leqslant \frac{e^{aX_2}}{e^{X_2}-1}.2\pi$

and this tends to zero as $X_2 \to \infty$ since $a < 1$.

On Γ_4 we have $z = -X_1 + it$ $(0 \leqslant t \leqslant 2\pi)$ and so

$$|f(z)| = \frac{|e^{a(-X_1+it)}|}{|e^{-X_1+it}+1|} \leqslant \frac{e^{-aX_1}}{1-e^{-X_1}}.$$

This gives $\left| \int_{\Gamma_4} f(z)dz \right| \leqslant \frac{e^{-aX_1}}{1-e^{-X_1}}.2\pi$

which tends to zero as $X_1 \to \infty$ because $a > 0$.

Thus the value of the infinite integral is proved.

By substituting $t = e^x$, we find

$$\int_{-\infty}^{\infty} \frac{e^{ax}}{e^x+1}\,dx = \int_0^{\infty} \frac{t^a}{t+1}\frac{dt}{t}.$$

This gives†

$$\int_0^\infty \frac{t^{a-1}}{t+1}\, dt = \frac{\pi}{\sin \pi a} \quad (0 < a < 1).$$

7. Summation of Certain Series

The functions $\cot \pi z$, $\operatorname{cosec} \pi z$ both have poles at 0, ± 1, $\pm 2, \ldots$ and so prove useful for summing series. If f is a function which is analytic at $z = n$, then $f(z) \operatorname{cosec} \pi z$ has a simple pole there with residue

$$\lim_{z \to n} (z-n)f(z) \operatorname{cosec} \pi z = \lim_{h \to 0} \frac{hf(n+h)}{\sin \pi(n+h)}$$

$$= \lim_{h \to 0} \frac{1}{\pi} \frac{\pi h}{(-1)^n \sin \pi h} f(n+h)$$

$$= \frac{(-1)^n f(n)}{\pi}.$$

Also $f(z) \cot \pi z = [f(z) \cos \pi z] \operatorname{cosec} \pi z$ has a simple pole at $z = n$ with residue $\dfrac{f(n)}{\pi}$.

Let S_N be the square with vertices $(N+\frac{1}{2})(\pm 1 \pm i)$ parametrized in the anticlockwise direction as in figure 16.

The contour S_N is chosen specifically because both $\cot \pi z$ and $\operatorname{cosec} \pi z$ are bounded on S_N. This requires some rather cumbersome calculations. First note that on the sides of S_N parallel to the real axis $z = x + iy$ where $|y| \geqslant \frac{1}{2}$, and on the other sides, $z = n + \frac{1}{2} + it$ where $n = \pm N$. If $z = n + \frac{1}{2} + it$ where $|y| \geqslant \frac{1}{2}$, then

---

† cf. W. Ledermann, *Integral Calculus*, pp. 64–67, where a proof of this result is given by real variable methods. It is of necessity very technical and this again illustrates the power of the theory of residues in those cases where it is applicable.

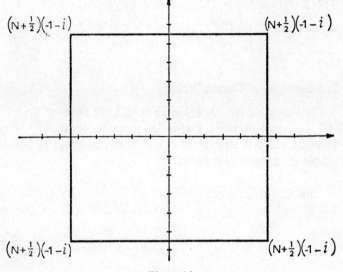

$\left(N+\tfrac{1}{2}\right)\!\left(-1-i\right)$ $\left(N+\tfrac{1}{2}\right)\!\left(-1-i\right)$

$\left(N+\tfrac{1}{2}\right)\!\left(-1-i\right)$ $\left(N+\tfrac{1}{2}\right)\!\left(-1-i\right)$

Figure 16

$$|\operatorname{cosec} \pi z| = (\tfrac{1}{2}|e^{i\pi z} - e^{-i\pi z}|)^{-1} \leqslant (\tfrac{1}{2}||e^{i\pi z}| - |e^{-i\pi z}||)^{-1}$$

$$= (\tfrac{1}{2}|e^{-\pi y} - e^{\pi y}|)^{-1} = (\sinh|\pi y|)^{-1} \leqslant \left(\sinh\frac{\pi}{2}\right)^{-1}.$$

Also $|\cot \pi z| = \left|\dfrac{\cos \pi z}{\sin \pi z}\right| = \left|\dfrac{e^{i\pi z} + e^{-i\pi z}}{e^{i\pi z} - e^{-i\pi z}}\right| \leqslant \left|\dfrac{|e^{i\pi z}| + |e^{-i\pi z}|}{|e^{i\pi z}| - |e^{-i\pi z}|}\right|$

$$= \left|\frac{e^{-\pi y} + e^{\pi y}}{e^{-\pi y} - e^{\pi y}}\right| = \coth|\pi y| \leqslant \coth\frac{\pi}{2}.$$

If $z = n + \tfrac{1}{2} + it$, then

$$|\operatorname{cosec} \pi z| = |\sin \pi z|^{-1} = |\cos i\pi t|^{-1} = (\cosh|\pi t|)^{-1} \leqslant 1,$$

and

$$|\cot \pi z| = |\tan it| = \left|\frac{1 - e^{-2t}}{1 + e^{-2t}}\right| \leqslant 1.$$

By Cauchy's Residue Theorem
$\int_{S_N} f(z) \cot \pi z \, dz$

$= 2\pi i \{$sum of residues of $f(z) \cot \pi z$ inside $S_N'\}$.

If $|f(z)| \leqslant \dfrac{A}{|z|^2}$ for $|z| \geqslant R$ where A, R are positive constants, then $\int_{S_N} f(z) \cot \pi z \, dz \to 0$ as $N \to \infty$. This follows because $|\cot \pi z|$ is bounded on S_N i.e. $|\cot \pi z| \leqslant M$ and so

$$\left| \int_{S_N} f(z) \cot \pi z \, dz \right| \leqslant \frac{A}{N^2} M(8N+4)$$

which tends to zero as $N \to \infty$.

Hence if $|f(z)| \leqslant \dfrac{A}{|z|^2}$ for $|z| \geqslant R$, then as $N \to \infty$, the sum of the residues of $f(z) \cot \pi z$ inside S_N tends to zero. Using the fact that if f is analytic at $z = n$ then $f(z) \cot \pi z$ has residue $\dfrac{f(n)}{\pi}$ there, this allows us to sum a series involving $f(n)$.

EXAMPLE. $f(z) = \dfrac{1}{z^2}$.

At an integer $n \neq 0$, $z^{-2} \cot \pi z$ has a simple pole with residue $1/(n^2 \pi)$. At the origin, as calculated on page 45, $z^{-2} \cot \pi z$ has a triple pole with residue $-\frac{1}{3}\pi$. Hence the sum of the residues of $f(z) \cot \pi z$ inside S_N is

$$\frac{1}{(-N)^2 \pi} + \dots + \frac{1}{(-1)^2 \pi} + \left(-\frac{1}{3}\pi \right) + \frac{1}{1^2 \pi} + \dots + \frac{1}{N^2 \pi}$$

$$= \frac{2}{\pi} \sum_{n=1}^{N} \frac{1}{n^2} - \frac{1}{3}\pi.$$

As $N \to \infty$, this tends to zero and so

$$\frac{2}{\pi} \sum_{n=1}^{\infty} \frac{1}{n^2} - \frac{1}{3}\pi = 0$$

$$\text{i.e. } \sum_{n=1}^{\infty} \frac{1}{n^2} = \frac{1}{6}\pi^2.$$

A similar calculation with cot πz replaced by cosec πz gives

$$\sum_{n=1}^{\infty} \frac{(-1)^{n-1}}{n^2} = \frac{1}{12}\pi^2.$$

EXERCISES ON CHAPTER THREE

1. Calculate the residues in the following cases:

   (i) $z^{-3}\sin^2 z$ $(z \neq 0)$, residue at $z = 0$.

   (ii) $\exp(1/z)$ $(z \neq 0)$, residue at $z = 0$.

   (iii) $e^z z^{-n-1}$ (n a positive integer, $z \neq 0$), residue at $z = 0$.

   (iv) $z^2(z^2 + a^2)^{-3}$ $(a > 0, z \neq \pm ia)$, residue at $z = ia$.

   (v) $(1 + z^2 + z^4)^{-1}$ $\left(z \neq \exp\left(\frac{r\pi i}{3}\right) \quad r = 1, 2, 4, 5\right)$, residue at $\exp\left(\frac{\pi i}{3}\right)$.

2. Show that $\displaystyle\int_0^{2\pi} \frac{\cos\theta}{2 - \cos\theta} d\theta = 2\pi\left(\frac{2}{\sqrt{3}-1}\right)$.

3. Show that $\displaystyle\int_0^{\pi} \frac{a}{a^2 + \sin^2 t} dt = \frac{\pi}{\sqrt{(1+a^2)}}$ $(a > 0)$. (Hint: substitute $\theta = 2t$.)

4. If C is the unit circle $z(t) = e^{it}$ $(0 \leqslant t \leqslant 2\pi)$, calculate by residues
$$\int_C e^z z^{-n-1} \, dz$$ where n is a positive integer. Hence show that
$$\int_0^{2\pi} \exp(\cos t) \cos(nt - \sin t) \, dt = \frac{2\pi}{n!}$$
$$\int_0^{2\pi} \exp(\cos t) \sin(nt - \sin t) \, dt = 0.$$

5. Evaluate $\displaystyle\int_0^\infty \frac{dx}{1 + x^2 + x^4}$.

6. Evaluate $\displaystyle\int_0^\infty \frac{\cos mx}{x^2 + a^2} \, dx$ $\quad (a > 0, \, m > 0)$.

7. Prove that $\displaystyle\int_0^\infty \frac{x^2}{(x^2 + a^2)^3} \, dx = \frac{\pi}{16a^3}$ $\quad (a > 0)$.

8. If $a > b > 0$, $m > 0$, prove that
$$\int_{-\infty}^\infty \frac{x^3 \sin mx}{(x^2 + a^2)(x^2 + b^2)} \, dx = \frac{\pi}{a^2 - b^2} (a^2 e^{-ma} - b^2 e^{-mb}).$$

9. Use the rectangle with vertices $-X_1$, X_2, $X_2 + \pi i$, $-X_1 + \pi i$ to show that $\displaystyle\int_{-\infty}^\infty \frac{e^{ax}}{\cosh x} \, dx = \frac{\pi}{\cos \frac{1}{2}\pi a}$ $\quad (-1 < a < 1)$.

10. Prove that $\displaystyle P\int_{-\infty}^\infty \frac{\cos x}{a^2 - x^2} \, dx = \frac{\pi \sin a}{a}$ $\quad (a > 0)$.

11. Show that $\displaystyle\sum_{n=1}^\infty \frac{(-1)^{n-1}}{n^2} = \frac{1}{12} \pi^2$.

12. Show that $\left(\dfrac{1}{\xi - z} + \dfrac{1}{z} \right) \cot \pi z$ has poles at every integer and at ξ.

Find the residues at these points when ξ is not an integer and in this case show that $\pi \cot \pi \xi = \dfrac{1}{\xi} + \displaystyle\sum_{n=1}^\infty \frac{2\xi}{\xi^2 - n^2}$.

CHAPTER FOUR

Analytic Continuation and Riemann Surfaces

---

1. Analytic Continuation

We now return to a topic discussed at the end of *Functions of a Complex Variable I* which will allow us to describe 'many-valued functions' in terms of (single-valued) functions. This is of interest when discussing contour integration because if f is analytic in a domain D and z_0 is a fixed point in D, then the integral of f along a contour γ from z_0 to an arbitrary point z in D depends in general on the choice of γ and so is in a sense 'many-valued'.

Recall† that if f and g are analytic functions defined in the same domain D and $f(z) = g(z)$ for all z in some non-empty open subset of D, then $f(z) = g(z)$ throughout the whole of D. It is this constraint on analytic functions, which forces two analytic functions to be equal everywhere in their joint domain of definition when they are only assumed equal on a small part, which leads to the results which we now explain.

Suppose that f_1 is defined in a domain D_1 and f_2 is defined in a domain D_2 where D_1 and D_2 overlap.

Under these conditions we say that f_2 is a *direct analytic continuation* of f_1 from D_1 to D_2. Of course if f_1 is analytic in D_1 and we are simply given the overlapping domain D_2, then we cannot be certain that a direct analytic continuation to D_2 exists. However if f_2 exists, then it is unique, for suppose that g is an alternative direct analytic continuation of f_1 to the domain D_2, then $g(z) = f_1(z) = f_2(z)$ for every point common to D_1 and D_2. But this set of points is a non-empty subset of

† *Functions of a Complex Variable I*, p. 62.

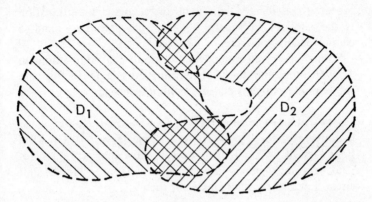

Figure 17

D_2 and is also open. (For if z lies in both D_1, D_2 then since D_1 is open there is an ε_1-neighbourhood of z lying completely in D_1. Similarly there is an ε_2-neighbourhood of z contained in D_2 and if ε is the smaller of ε_1, ε_2 then the ε-neighbourhood of z lies in the overlap of D_1 and D_2 which shows that this overlap is open.) Hence $g(z) = f_2(z)$ throughout D_2.

The notion of direct analytic continuation is most often used when D_2 contains D_1. Here we begin with an analytic function f_1 in D_1 and try to find an analytic function f_2 defined on the larger domain D_2 which equals f_1 on D_1. This idea of extending the domain on which an analytic function is defined was discussed in *Functions of a Complex Variable I*, pages 60-62.

EXAMPLE 1. $f(z) = \sum_{n=0}^{\infty}(-1)^n z^{2n} \quad |z| < 1.$

The function $(1+z^2)^{-1}$ is defined and analytic for $z \neq \pm i$ and equals $\sum_{n=0}^{\infty}(-1)^n z^{2n}$ for $|z| < 1$. Hence $(1+z^2)^{-1}$ is a direct analytic continuation to the domain consisting of all points

65

except $\pm i$. Evidently there is no direct analytic continuation to the whole plane because $(1+z^2)^{-1}$ has poles at $\pm i$ and so cannot be analytic there.

Sometimes, given an analytic function f defined in a domain D, we cannot continue f analytically outside D. In this case the boundary of D is called a *natural boundary*.

EXAMPLE 2. The series $f(z) = 1 + z + z^2 + z^4 + \ldots + z^{2^n} + \ldots$ is convergent for $|z| > 1$. The unit circle $|z| = 1$ is a natural boundary. If $w^{2^m} = 1$, then we can show that $f(z)$ does not tend to a finite limit as z approaches w from inside the unit circle. Let $z = rw$ where $0 < r < 1$, then

$$f(z) = 1 + z + z^2 + \ldots + z^{2^{m-1}} + \sum_{n=m}^{\infty} z^{2^n}$$
$$= f_1(z) + f_2(z).$$

We have $\lim_{r \to 1} f_1(rw) = 1 + w + w^2 + \ldots + w^{2^{m-1}}$. But since $w^{2^m} = 1$, the series $f_2(rw) = \sum_{n=m}^{\infty} r^{2^n}$ is a series of real, positive terms for $0 < r < 1$. Hence $f_2(rw) > \sum_{n=m}^{m+N} r^{2^n}$. But $\sum_{n=m}^{m+N} r^{2^n} \to N+1$ and so for some $\varepsilon > 0$, if $1 - \varepsilon < r < 1$ then $\sum_{n=m}^{m+N} r^{2^n} > \frac{1}{2}N$. This gives $f_2(rw) > \frac{1}{2}N$ and since N is arbitrary, $f_2(rw) \to +\infty$ as $r \to 1$.

Thus f cannot be analytically continued into any domain containing w where $w^{2^m} = 1$. But if a domain D_2 crosses the circle $|z| = 1$, then it includes a segment of the circle. The roots of $z^{2^m} = 1$ are $\exp\left(\dfrac{2\pi i q}{2^m}\right)$ where $q = 1, \ldots, 2^m$. These are spaced at equal intervals around the unit circle. By choosing m large enough, some point w where $w^{2^m} = 1$ lies in the segment of the circle inside D_2. Thus f cannot be analytically continued across $|z| = 1$.

In some cases the process of direct analytic continuation may be repeated. Given a function f_1 defined in a domain D_1, we may find a direct analytic continuation f_2 to a domain D_2 where D_1 and D_2 overlap. Then we may find a direct analytic continuation f_3 of the function f_2 to a domain D_3 where D_2 and D_3 overlap. After a finite number of steps we find a direct analytic continuation f_n of f_{n-1} from D_{n-1} to D_n. In this case, f_n is called an *indirect* analytic continuation to the domain D_n of the function f_1 defined in D_1. We refer to both direct and indirect analytic continuations simply as 'analytic continuations'. Any two analytic continuations of a given function are evidently analytic continuations of each other.

The theory of indirect analytic continuation is much more complicated than direct continuation. The main problem is that it need no longer be unique. This is because we might use a different sequence of domains linking D_1 to D_n.

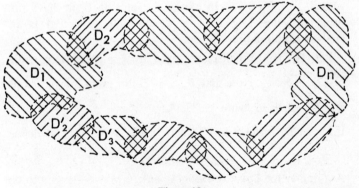

Figure 18

For example we might eventually return to the original domain and have $D_1 = D_n$, but find the indirect continuation f_n different from the original function f_1. We define the *complete analytic function* to consist of the original function and all its possible analytic continuations. In the case where we have different analytic continuations to some domain, the complete analytic function is called *multiform*, otherwise it is called *uniform*. Examples 1, 2 are uniform.

If no analytic continuation can be defined at a point z_0, then z_0 is said to be a *singularity* of the complete analytic function. In example 1, the points $z = \pm i$ are singularities and in example 2 all the points $|z| \geq 1$ are singularities.

Note that a multiform complete analytic function is in a sense 'many-valued', but we have formulated it as a collection of (single-valued) functions. Two functions in the collection may have different values in the same domain, but they are analytic continuations of each other.

EXAMPLE 3. The logarithm is multiform. For any integer k we can define $\log_k z$ in the cut-plane by

$$\log_k z = \log |z| + i\,(\arg z + 2\pi k)$$

where $-\pi < \arg z < \pi$. In particular, for $k = 0$, we have the principal value $\text{Log } z = \log_0 z$. We will show $\log_k$ is an analytic continuation of Log.

Let D_n be the half-plane given by $z = re^{i\theta}$ where $r > 0$, $(n-2)\dfrac{\pi}{2} < \theta < \dfrac{n\pi}{2}$. Note that $D_{n+4} = D_n$ for every integer n and D_1, D_2 are as in figure 19.

If z is in D_n, write $z = re^{i\theta}$ where $(n-2)\dfrac{\pi}{2} < \theta < \dfrac{n\pi}{2}$ and define

$$f_n(z) = \log r + i\theta.$$

Arguing as for Log z in the cut-plane, $f_n(z)$ may be seen to be analytic in the domain D_n. If $z = re^{i\theta}$ is in D_n and D_{n+1}, then $f_n(z) = f_{n+1}(z)$ and so f_{n+1} is the direct analytic continuation of f_n from D_n to D_{n+1}. By induction, f_m is an analytic continuation of f_n from D_n to D_m for any m and n. In particular, in the domain $D_{n+4} = D_n$, we see that $f_{n+4}(z) = f_n(z) + 2\pi i$ is an analytic continuation of $f_n(z)$. The function $\log_k$ defined in the

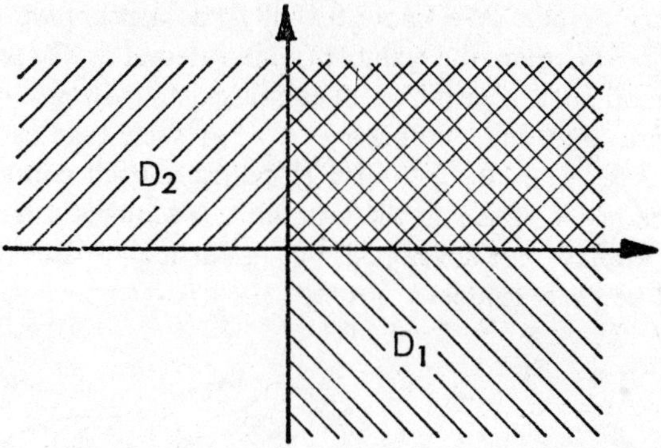

Figure 19

cut-plane coincides in D_1 with the function f_{4k+1}. Thus f_{4k+1} is trivially a direct analytic continuation of $\log_k$. If we start with $\text{Log} = \log_0$ in the cut-plane, we find a chain of analytic continuations, f_1 in D_1, f_2 in $D_2, \ldots, f_{4k+1}$ in $D_{4k+1} = D_1$ and finally $\log_k$ in the cut-plane, showing that $\log_k$ is an analytic continuation of Log in the cut-plane.

Note that f_{k+3} is defined in $D_3 \Big($ the half-plane $z = re^{i\theta}, r > 0,$ $\dfrac{\pi}{2} < \theta < \dfrac{3\pi}{2} \Big)$ and D_3 includes all the points on the negative real axis except the origin. Hence the analytic continuation f_{k+3} of Log is defined on the negative real axis except the origin. Thus the only singularity of the complete analytic function can be at the origin. By analytically continuing via a set of domains round the origin we obtain different analytic continuations. In general a singularity with this property is called a *branch point*.

We now consider multiform examples which appear naturally in contour integration.

If f is analytic in a domain D, fix a point z_0 in D and consider the integral of f along a contour γ in D from z_0 to an

69

arbitrary point z. We know that if f has a primitive F in D (i.e. $F' = f$), then the value of this integral is $F(z) - F(z_0)$. In general such a primitive does not exist. However we may subdivide into subcontours $\gamma_1, \ldots, \gamma_n$ such that each subcontour γ_r lies in an open disc D_r which is itself contained in D. (The proof of this in the general case requires a technique which we have not developed, but in particular cases its truth should be fairly evident.)

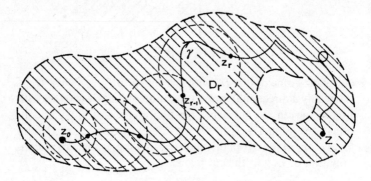

Figure 20

Now in a disc an analytic function *does*† have a primitive which is unique up to an additive constant. Let F_r be a primitive for f in D_r ($r = 1, \ldots, n$). By definition $F_r' = f$ in D_r and $F_{r+1}' = f$ in D_{r+1} and so $F_r' - F_{r+1}' = 0$ in the overlap. But the overlap of two circles is a domain and so

$$F_{r+1}(z) = F_r(z) + \text{constant}$$

for all z in both D_r and D_{r+1}. By adding a suitable constant to each of $F_2, F_3, \ldots, F_n$ in turn, we may suppose that $F_{r+1} = F_r$ in the overlap of D_r and D_{r+1} for $r = 1, 2, \ldots, n-1$. This yields an example of analytic continuation.

† *Functions of a Complex Variable I*, p. 47.

Let the initial and final points of γ_r be z_{r-1}, z_r then by the Fundamental Theorem of Contour Integration,

$$\int_{\gamma_r} f(z)dz = F_r(z_r) - F_r(z_{r-1}). \qquad (1)$$

Since z_{r-1} lies in the overlap of D_{r-1} and D_r, we have $F_{r-1}(z_{r-1}) = F_r(z_{r-1})$. Adding up the integrals along the subcontours and cancelling $F_{r-1}(z_{r-1}) - F_r(z_{r-1})$ for $r = 2, \ldots, n$, we find

$$\int_{\gamma} f(z)dz = F_n(z_n) - F_1(z_0). \qquad (2)$$

Of course if f had a primitive F throughout D then in particular $F_1' = F'$ in D_1. Adding a constant if necessary, we may suppose that $F_1 = F$ restricted to D_1. By successive direct analytic continuations, we then find that $F_r = F$ restricted to D_r for $r = 1, \ldots, n$ and so (2) reduces to the Fundamental Theorem $\int_{\gamma} f(z)dz = F(z_n) - F(z_0)$.

However, if f has an isolated singularity in D with non-zero residue ρ, then selecting a closed Jordan contour γ in D winding once anticlockwise round this singularity, we find

$$\int_{\gamma} f(z)dz = 2\pi i\rho. \qquad (3)$$

Since γ is closed, $z_0 = z_n$ and from (2), (3), $F_n(z_0) = F_1(z_0) + 2\pi i\rho$. Hence F_1, F_2 are *not* equal and we have an example which is multiform. The isolated singularity of f gives a branch point of the complete analytic function found by continuing the primitive F_1.

2. Riemann Surfaces

The notion of analytic continuation explained in the last section is quite difficult for the beginner to grasp. In particular it is difficult to visualize an overall picture of what is going on. This total view of the situation is best described by using the idea of a 'Riemann surface'. We will illustrate this concept by two particular examples, first considering the case of the logarithm.

If $z = e^w$, then all the solutions for w in terms of z (where $z \neq 0$) are given by $w = \log|z| + i(\arg z + 2\pi k)$ where $-\pi < \arg z \leqslant \pi$, and k is an integer. Restricting ourselves to the principal value given by $k = 0$ in the cut-plane, we have an analytic function and in the last section we saw that we could recover all the other values by analytic continuation. Each time we pass round the origin in the anti-clockwise direction the value of $w = \log z$ is increased by $2\pi i$.

We now describe another method of looking at this phenomenon by introducing a Riemann surface. It will have the advantage that we obtain a single-valued function which takes all the values of the logarithm but this function will now be defined on the Riemann surface and not on the complex plane.

Consider the complex plane to be covered by an infinite number of superimposed transparent sheets (each sheet covers the whole plane). From every sheet remove the origin and imagine a cut being made along the negative real axis in such a way that this axis is considered to be affixed to the upper part of the cut. Now smoothly join the negative real axis of the upper part of the cut on each sheet to the lower part of the cut on the sheet above. If we mark a point on one of the sheets and imagine it to move over the cut in the anti-clockwise direction then, because of the smooth join, we suppose that it moves on to the next sheet above. This means that if the superimposed sheets were pulled apart and viewed from the side, then the system would look rather like an infinite winding staircase. This system of sheets is called the *Riemann surface* of the logarithm.

Looking at the Riemann surface from above, since the sheets are transparent, marking a point on one of them represents a non-zero complex number. However, given two real numbers r, θ where $r > 0$ and $(2k-1)\pi < \theta \leqslant (2k+1)\pi$, then by numbering the sheets in ascending order we can suppose that the pair of numbers r, θ gives the point on the kth sheet which represents the complex number $re^{i\theta}$. Thus the Riemann surface may be considered to have the advantage of distinguishing between the symbols $re^{i(\theta+2\pi k)}$, $k = 0, \pm 1, \pm 2, \ldots$, which are equal as complex numbers, but lie vertically above one another on different sheets of the Riemann surface.†

Define the logarithm on the Riemann surface by

$$\log P = \log |z| + i(\arg z + 2\pi k)$$

where P is the point on the kth sheet representing the complex number z. Alternatively, if $z = re^{i\theta}$ where $r > 0$, $(2k-1)\pi < \theta \leqslant (2k+1)\pi$, then

$$\log P = \log r + i\theta.$$

Note that the logarithm is a single-valued function on the Riemann surface. It is also continuous, in the intuitive sense that as a point P tends to P_0, then $\log P$ tends to $\log P_0$ (even when P moves over the cut from one sheet to the next).

We can now begin to see what happens when we analytically continue some analytic, single-valued choice of the logarithm. To do this we just look at the corresponding situation on the Riemann surface.

We first remark that if we are given an analytic function f defined in a domain D where $f(z)$ is always a logarithm of z, then this gives us a rule to choose a 'domain' on the Riemann surface which corresponds to the domain D in the complex plane. This is because f is a choice of logarithm and so the imaginary part of $f(z)$ is a particular choice θ for an argument of z. For each point z in D we then select the point P on the Riemann surface which represents $z = re^{i\theta}$ on the kth sheet where $(2k-1)\pi < \theta \leqslant (2k+1)\pi$. (This construction does no

† This may be considered in three-dimensional space as the surface given parametrically by $(r \cos \theta, r \sin \theta, \theta)$ where $r > 0$. The kth sheet is given by $(2k-1)\pi < \theta \leqslant (2k+1)\pi$.

more than choose the point P on the appropriate sheet according to the actual value of θ as previously described.) Since f is analytic, its imaginary part is continuous i.e. θ depends continuously on z and if we imagine z moving continuously about in D, the corresponding point P moves continuously about in the corresponding 'domain'.

Now suppose that we analytically continue f outside the domain D. As we move successively from one domain to an overlapping one, we imagine the corresponding movement on the Riemann surface. If the set of domains used wanders round the origin then the corresponding set of domains on the Riemann surface passes round the 'winding staircase', possibly moving up or down to a different level. This possibility of arriving at another level gives a clear geometrical picture of why we can get different analytic continuations into a given domain by choosing alternative routes. The notion of analytic continuation described in section 1 is just a 'flattened version' in the complex plane of what is happening on the Riemann surface.

We can represent other 'many-valued functions' as single-valued functions on Riemann surfaces. In general, an 'n-valued function' requires n sheets. We illustrate this by considering $z^{\frac{1}{2}}$. This requires two sheets each with the origin removed and cut along the negative real axis. If $z = re^{i\theta}$ where $r > 0$, $-\pi < \theta \leqslant \pi$, then choose $z^{\frac{1}{2}} = r^{\frac{1}{2}}e^{\frac{1}{2}i\theta}$ on the first sheet and $z^{\frac{1}{2}} = r^{\frac{1}{2}}e^{\frac{1}{2}i(\theta + 2\pi)} = -r^{\frac{1}{2}}e^{\frac{1}{2}i\theta}$ on the second. As a point moves over the cut in the anti-clockwise direction, it passes from the first sheet to the second and after a complete circuit round the origin again, when it crosses the cut again, it passes from the second sheet back to the first. The Riemann surface is found by taking the two sheets in figure 21 and joining together the sides of the negative real axis marked '$+$' and those marked '$-$'.

This construction can only be performed in an idealized situation since it is not possible to physically cut two actual

Sheet I   Sheet II

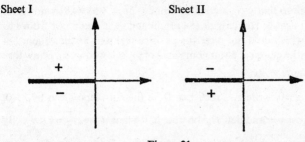

Figure 21

sheets and glue them together in the manner described without
unwanted self-intersections. (After glueing '+' to '+', for
example, it is not possible to fix '−' to '−' without passing
through the glued sheets.) However by a stretch of the imagina-
tion using figure 21 it should be possible to visualize the
idealized concept. This brings us to a fitting point to end the
discussion as the mind grapples with an idea beyond the
confines of three dimensional existence.

EXERCISES ON CHAPTER FOUR

Find analytic continuations of the following power series:

1. $\displaystyle\sum_{n=1}^{\infty} (-1)^n z^n \quad |z| < 1$

2. $\displaystyle\sum_{n=0}^{\infty} z^{3n} \quad |z| < 1.$

3. $\displaystyle\sum_{n=0}^{\infty} 3n z^{3n-1} \quad |z| < 1.$

4. What happens when we look for the analytic continuations of
$\displaystyle\sum_{n=1}^{\infty} (-1)^{n-1}(z^n/n)$ outside the disc $|z| < 1$?

5. Show that $|z| = 1$ is a natural boundary for $\displaystyle\sum_{n=0}^{\infty} z^{n!} \quad |z| < 1.$

6. Suppose that γ_1 is a contour from $-i$ to i which does not meet the negative real axis or the origin and γ_2 is a contour from i to $-i$ which does not meet the positive real axis or the origin. Let γ be the closed contour composed of γ_1 followed by γ_2. Show that
$$\int_\gamma 1/z \, dz = 2\pi i.$$

7. For $z \neq 0$, write $z = re^{i\theta}$. Let D be the cut-plane given by $r > 0$, $-\pi < \theta < \pi$ and let D_n be the half-plane $(n-2)\dfrac{\pi}{2} < \theta < n\dfrac{\pi}{2}$. By successive direct analytic continuations from D to D_1, from D_1 to D_2, from D_2 to D_3, from D_3 to D_4, from D_4 to $D_5 = D_1$, and from D_1 back to D, show that $-z^{\frac{1}{2}}$ is an indirect analytic continuation of $z^{\frac{1}{2}}$ in D (where $z^{\frac{1}{2}} = r^{\frac{1}{2}}e^{\frac{1}{2}}$, $r > 0$, $-\pi < \theta < \pi$).

Describe the Riemann surfaces for the following 'many-valued functions':

8. $z^{\frac{1}{2}}$. 9. $z^{\frac{1}{n}}$ for a positive integer n. 10. $(z-1)^{-\frac{1}{2}}$.

Solutions to Exercises

Chapter One

1. (i) $w_1(t) = e^t$ $(-1 \leqslant t \leqslant 1)$, $w_2(t) = e^{t(1+i)}$ $(-1 \leqslant t \leqslant 1)$, angle between curves is $\arg w_1'(0) - \arg w_2'(0) = \arg 1 - \arg(1+i) = \dfrac{\pi}{4}$, similarly for (ii), (iii).

2. $w_1(t) = t^n$ $(0 \leqslant t \leqslant 1)$, $w_2(t) = t^n e^{in\alpha}$ $(0 \leqslant t \leqslant 1)$. The first has track given by $y = 0$, $0 \leqslant x \leqslant 1$, the second is the line segment $y = x \tan n\alpha$ from $(0, 0)$ to $(\cos n\alpha, \sin n\alpha)$. These two lines are at an angle $n\alpha$.

3. $c(x^2 + y^2) + x = 0$, circles touching imaginary axis at the origin, $k(x^2 + y^2) + y = 0$, circles touching real axis at the origin.

5. $ax^3 - 3dx^2y - 3axy^2 + dy^3$. $f(z) = (a + id)z^3 + ik$, (k real).

6. $(2 + 2i)\sin z + (1 + 2i)z^2 + ik$, ($k$ real).

Chapter Two

1. $z^{-5} + z^{-2} + \dfrac{z}{2!} + \ldots + \dfrac{z^{3n-5}}{n!} + \ldots$ $(z \neq 0)$ pole of order 5.

2. $\dfrac{1}{2a(z-a)} - \dfrac{1}{4a^2} + \ldots + \dfrac{(-1)^{n+1}(z-a)^n}{(2a)^{n+2}} + \ldots$ $(0 < |z - a| < 2a)$ simple pole.

3. $\dfrac{1}{z} - \dfrac{1}{2!z^3} + \ldots + \dfrac{(-1)^n}{(2n)!z^{2n+1}} + \ldots$ $(z \neq 0)$ essential singularity.

4. $(1 - z)e^{\frac{1}{z}} = \left(1 + \dfrac{1}{z} + \dfrac{1}{2!z^2} + \ldots + \dfrac{1}{n!z^n} + \ldots \right)$

$$- \left(z + 1 + \dfrac{1}{2!z} + \ldots + \dfrac{1}{(n+1)!z^n} + \ldots \right)$$

$$= -z + \tfrac{1}{2}z^{-1} + \ldots + \dfrac{n \ z^{-n}}{(n+1)!} + \ldots$$

$(z \neq 0)$ essential singularity.

5. $\dfrac{1}{12z} - \dfrac{2z}{6!} + \ldots + \dfrac{(-1)^n 2 z^{2n-5}}{(2n)!} + \ldots$ ($z \neq 0$) simple pole.

6. $-(z-1)^{-1} - 1 - (z-1) - \ldots - (z-1)^n - \ldots$ ($0 < |z-1| < 1$) simple pole.

7. (a) pole of order 3 $\left(\text{since } \lim_{z \to 0} z^3 \, \dfrac{e^z}{z \sin^2 z} = 1\right)$

   (b) essential singularity (since $n\pi$ is a singularity for every integer n).

8. (a) simple pole $\left(\text{since } \lim_{z \to 0} z \, \dfrac{z}{1 - \cos z} = \lim_{z \to 0} \dfrac{4(\tfrac{1}{2}z)^2}{2 \sin^2(\tfrac{1}{2}z)} = 2\right)$.

   (b) essential singularity (since $(2n + \tfrac{1}{2})\pi$ is a singularity for every integer n).

9. (a) isolated essential singularity.   (b) pole of order 2.

10. (a) essential singularity.   (b) removable singularity.

11. (a) pole of order 3.   (b) essential singularity.

12. Use $g(z) = e^z$, $f(z) = -\alpha z^n$ in Rouché's Theorem.

Chapter Three

1. (i) 1   (ii) 1   (iii) $\dfrac{1}{n!}$ (iv) $\dfrac{-i}{16a^3}$ (v) $(-3 + i\sqrt{3})^{-1}$.

5. $\dfrac{\pi\sqrt{3}}{6}$.   6. $\dfrac{\pi e^{-ma}}{2a}$.

12. residue of $\left(\dfrac{1}{\xi - z} + \dfrac{1}{z}\right) \cot \pi z$ at $n \neq 0$ is $\left(\dfrac{1}{\xi - n} + \dfrac{1}{n}\right)/\pi$, at the origin it is $1/\pi\xi$ and at ξ it is $\cot \pi\xi$.

Chapter Four

1. $(1+z)^{-1}$ $z \neq -1$ 2. $(1 - z^3)^{-1}$ $z \neq 1$, $e^{2\pi i/3}$, $e^{4\pi i/3}$

3. $3z^2(1 - z^3)^{-2}$ $z \neq 1$, $e^{2\pi i/3}$, $e^{4\pi i/3}$ (hint: differentiate $(1 - z^3)^{-1}$)

4. $\sum (-1)^{n-1}(z/n) = \text{Log}(1+z)$ $|z| < 1$. Indirect analytic continuation gives all the values of the logarithm of $1 + z$ (where $z \neq -1$).

78

5. If $z_0^m = 1$, then $\lim_{z \to z_0} \sum z^{n!}$ does not exist by a proof analogous to that for $\sum z^{2^n}$ given in the text. Any domain crossing $|z| = 1$ contains such a point.

6. Log $z = \log|z| + i$ arg z $(-\pi < \text{arg } z < \pi)$ is analytic in the cut-plane with the negative real axis including the origin removed. $\frac{d}{dz}\left(\text{Log } z\right) = 1/z$. $\int_{\gamma_1} 1/z\,dz = \text{Log } i - \text{Log}(-i) = \pi i$. Similarly $\log_* z = \log|z| + i$ arg$_* z$ $(0 < \text{arg}_* z < 2\pi)$ is analytic in the cut-plane with the positive real axis and the origin removed. Here $\frac{d}{dz}(\log_* z) = 1/z$ and $\int_{\gamma_2} 1/z\,dz = \log_*(-i) - \log_* i = \pi i$. Hence $\int_\gamma 1/z\,dz = \pi i + \pi i = 2\pi i$. (Remark: Any closed contour γ not passing through the origin satisfies $\int_\gamma 1/z\,dz = 2n\pi i$ where n is an integer. The integer n is the number of times γ winds round the origin. Try to visualize this by considering the situation on the Riemann surface for the logarithm.)

7. $z^{\frac{1}{2}} = r^{\frac{1}{2}}e^{\frac{1}{2}i\theta}$ $\left(-\frac{\pi}{2} < \theta < \frac{\pi}{2}\right)$ in D_1, continuation $r^{\frac{1}{2}}e^{\frac{1}{2}i\theta}$ $(0 < \theta < \pi)$ in D_2, $r^{\frac{1}{2}}e^{\frac{1}{2}i\theta}$ $\left(\frac{\pi}{2} < \theta < \frac{3\pi}{2}\right)$ in D_3, $r^{\frac{1}{2}}e^{\frac{1}{2}i\theta}$ $(\pi < \theta < 2\pi)$ in D_4, $r^{\frac{1}{2}}e^{\frac{1}{2}i\theta}$ $\left(\frac{3\pi}{2} < \theta < \frac{5\pi}{2}\right)$ in $D_5 = D_1$. Replacing θ by $\theta + 2\pi$, this may be re-written as $r^{\frac{1}{2}}e^{\frac{1}{2}i\theta + i\pi}$ $\left(-\frac{\pi}{2} < \theta < \frac{\pi}{2}\right)$ in D_1. But $r^{\frac{1}{2}}e^{\frac{1}{2}i\theta + i\pi} = -r^{\frac{1}{2}}e^{\frac{1}{2}i\theta}$ which has continuation $-z^{\frac{1}{2}} = -r^{\frac{1}{2}}e^{\frac{1}{2}i\theta}$ $(-\pi < \theta < \pi)$ in D.

8, 9. Take n sheets cut along the negative real axis and join the upper part of the cut on sheet r to the lower part of the cut on sheet $r+1$ for $r = 1, 2, \ldots, n-1$ and join the upper part on sheet n to the lower part on sheet 1.

10. As for the Riemann surface of $z^{\frac{1}{2}}$ but with the cut on each sheet along the negative real axis through the origin as far as $z = 1$.

INDEX TO PARTS I AND II